Springer Series in Wood Science

Editor: T. E. Timell

M. H. Zimmermann
Xylem Structure and the Ascent of Sap (1983)

J. F. Siau
Transport Processes in Wood (1984)

R. R. Archer
Growth Stresses and Strains in Trees (1986)

W. E. Hillis
Heartwood and Tree Exudates (1987)

S. Carlquist
Comparative Wood Anatomy (1988)

L. W. Roberts / P. B. Gahan / R. Aloni
Vascular Differentiation and Plant Growth Regulators (1988)

C. Skaar
Wood-Water Relations (1988)

J. M. Harris
Spiral Grain and Wave Phenomena in Wood Formation (1989)

B. J. Zobel / J. P. van Buijtenen
Wood Variation (1989)

P. Hakkila
Utilization of Residual Forest Biomass (1989)

J. W. Rowe
Natural Products of Woody Plants (1989)

K.-E. L. Eriksson / R. A. Blanchette / P. Ander
Microbial and Enzymatic Degradation of Wood and Wood Components (1990)

R. A. Blanchette / A. R. Biggs
Defense Mechanisms of Woody Plants Against Fungi (1992)

S. Y. Lin / C. W. Dence
Methods in Lignin Chemistry (1992)

G. Torgovnikov
Dielectric Properties of Wood and Wood-Based Materials (1993)

G. I. Torgovnikov

Dielectric Properties of Wood and Wood-Based Materials

With 82 Figures

Springer-Verlag
Berlin Heidelberg New York
London Paris Tokyo
Hong Kong Barcelona Budapest

Grigoriy I. Torgovnikov
Central Scientific Research Institute of Forest Industry,
Mechanization and Power Engineering (TsNIIME)
141400 Khimki, Moscow Region, Russia

Series Editor:

T. E. Timell
State University of New York, College of Environmental, Science and Foresty,
Syracuse, NY 13210, USA

Cover: Transverse section of *Pinus lambertiana* wood. Courtesy of Dr. Carl de Zeeuw,
SUNY College of Environmental Science and Forestry, Syracuse, New York

ISBN-13:978-3-642-77455-3 e-ISBN-13:978-3-642-77453-9
DOI: 10.1007/978-3-642-77453-9

Softcover reprint of the hardcover 1st edition 1993

31/3020 - 5 4 3 2 1 0 - Printed on acid-free paper

Preface

There is an increasing interest in the study of the dielectric properties of wood as a result of the increased utilization of high frequency and microwave techniques for the intensification of processes such as heating, drying, glueing, and improvement of the quality of wood and wood-based materials. There is also a need for designing modern instruments for the inspection of quantitative and qualitative characteristics of wood-based materials and finished products. Development of technologies using electric fields requires a profound knowledge of the dielectric properties of wood.

The electric characteristics that wood reveals under the influence of a direct current and of a low-frequency alternating current (20–50 Hz) have been studied since the beginning of the 20th century. Exploration of wood properties elicited by the action of high frequency alternating current (or AC) began somewhat later, in 1930–1934, associated with the development of high-frequency sources of power by the beginning of the 1930s. One of the first areas of industrial use of such sources was wood drying processes. With the progress of the microwave technique in the 1950s, the industry received microwave instruments for measuring the moisture content and thickness of a material and for the detection of defects and also obtained powerful generators suitable for the dielectric heating of wood and wood-based materials.

The scientific workers of different countries, such as K. Kröner, L. Pungs, W. Trapp, C. Skaar, A.R. von Hippel, A.G.H. Dietz, R.W. Peterson, R.F.S. Hearmon, J.N. Burcham, T. Uyemura, S.T. Lee, W.L. James, J. Tsutsumi, W.R. Tinga, M. Norimoto, K.P. Mikhailovskaja, I. Makoviny, and others have contributed much to the study of the dielectric properties of wood. The important research contributions of M. Norimoto deserve special mention.

In this book an attempt has been made to systematize the knowledge accumulated regarding the dielectric properties of wood. The data in this book on the dielectric properties of wood and on the influence of the species, density, moisture content, anisotropy, and temperature of the wood, and of the electromagnetic field frequency represent current knowledge in this field.

The book comprises nine chapters. Chapter 1 deals with the interactions between the electromagnetic field and the wood and with the kinds of polarization that take place in the wood. This chapter also includes data on wood dielectric parameters and their ratio and some mathematical dependencies reflecting interactions between the electromagnetic fields and the material. Chapter 2 considers wood as a multicomponent dielectric material and analyses the dielectric properties of its components. In Chapter 3, the methods of measurement of the dielectric characteristics of wood are described . Chapters 4 and 5 examine the dielectric properties of dry and moist wood and the influence of wood species, density, and anisotropy, as well as of temperature and frequency on these properties. Chapter 6 is concerned with the influence of mechanical and chemical treatments on the dielectric properties of wood. This chapter also deals with the influence of gamma-radiation on the dielectric characteristics of wood. The dielectric properties of bark are discussed in Chapter 7. Chapter 8 is concerned with the dielectric properties of cellulose, paper, board, fiberboards, particleboards, and other wood-based materials. Chapter 9 includes some recommendations on the choice and the use of the dielectric parameters in calculations.

Appendices comprise reference data on the dielectric characteristics of wood and wood-based materials in a wide range of frequencies, temperatures, and moisture content values.

I hope that this book will prove to be useful for scientists and for design and industrial engineers, who create new production processes, develop high-frequency and microwave equipment for the treatment of wood and wood-based materials, and measure different parameters of the materials. I also hope that this book will promote the use of electrotechnology in the woodworking industry. I believe that this book will be useful for students who are interested in scientific wood problems and in the theories of dielectric materials.

I express my gratitude to my colleagues T.V. Minakova and N.V. Arkhipova of the Central Scientific Research Institute of Forest Industry Mechanization and Power Engineering (TsNIIME), Khimki, Moscow Region, Russia, for their help in the research and in preparation of the manuscript. I am very grateful to Dr. T.E. Timell for editing this book, and to Prof. C. Skaar for making scientific articles available to me.

Moscow, November 1992 Grigoriy Torgovnikov

Contents

List of Symbols

a, b	– proportionality factors
c	– velocity of light
C	– capacitance
c.w.s.	– cell wall substance
d	– penetration depth
DC	– dielectric constant
e	– natural logarithmic base
E	– electric field strength
f	– frequency
g	– relative volume of pores in paper
$i = \sqrt{-1}$	– imaginary unit
J	– current density
k_1, k_2, k_3	– coefficients which characterize anisotropy of dielectric constant, loss tangent, and loss factor values of wood with the vector $\bar{E}$ oriented parallel to and perpendicular to the fiber direction ($k_1 = \varepsilon'_{\parallel}/\varepsilon'_{\perp}$, $k_2 = tg\delta_{\parallel}/tg\delta_{\perp}$, $k_3 = \varepsilon'_{\parallel}/\varepsilon''_{\perp}$)
$k_{1(R/T)}$ $k_{2(R/T)}$	– coefficients which characterize anisotropy of dielectric constant and loss tangent values with vector $\bar{E}$ oriented in radial and tangential directions
$k^t_{\varepsilon'}$, $k^t_{tg\delta}$	– temperature coefficients of dielectric constant and loss tangent values
K	– Kelvin
l	– length
L	– longitudinal direction
m	– mass, meter
N	– power value
P	– polarization
Q	– quality factor
R	– radial direction, electric resistance
s	– degree of impregnation
t	– temperature, °C
$tg\delta$	– loss tangent
$tg\delta_L$, $tg\delta_R$, $tg\delta_T$	– loss tangent values with the vector $\bar{E}$ orientation in longitudinal, radial, and tangential directions
$tg\delta_{LR}$, $tg\delta_{LT}$, $tg\delta_{RT}$	– average values of loss tangent at the change of vector $\bar{E}$ orientation from longitudinal to radial, from longitudinal to tangential, and from radial to tangential directions
T	– absolute temperature, tangential direction

u – voltage
v – velocity
V – volume, relative volume content, volts
w – relative quantity of nonfreezing moisture
W – moisture content
W' – energy
x – coordinate
z – resistance
Z – impedance
α – angle, relaxation time distribution constant, attenuation constant
β – phase constant
γ_1 – propagation constant
δ – loss angle
Δ – attenuation distance
ΔC – change of capacitance value
ΔE – activation energy
$\Delta 1$ – distance
ε_0 – permittivity of free space
ε^* – complex dielectric constant
ε' – relative dielectric constant
ε'_L, ε'_R, ε'_T – dielectric constant values at the vector $\bar{E}$ orientation in longitudinal, radial, and tangential directions
ε'_∞ – limiting high-frequency dielectric constant (optic)
ε'_s – limiting low-frequency dielectric constant (static)
ε'_{LR}, ε'_{LT}, ε'_{RT} – average values of dielectric constant at the changes of vector $\bar{E}$ orientation from longitudinal to radial, from longitudinal to tangential, and from radial to tangential directions
ε'' – loss factor
ε''_L, ε''_R, ε''_T – loss factor values with the vector $\bar{E}$ oriented in longitudinal, radial, and tangential directions
ε'_{LR}, ε'_{LT}, ε'_{RT} – average values of loss factor at the changes of vector $\bar{E}$ orientation from longitudinal to radial, from longitudinal to tangential, and from radial to tangential directions
λ – wavelength
λ_0 – wavelength in the air
λ_b – wavelength in the waveguide
μ_1 – compression degree
π – pi (3.14)
ρ – density
ρ_0 – density value of oven-dry material
σ – specific conductance
σ' – active component of specific conductance
σ'' – reactive component of specific conductance
Σ – sum, total value

τ – time
τ_0 – time constant of dielectric relaxation
φ – angle, function, phase displacement angle
χ – dielectric susceptibility
ψ – angle, function
ω – angular frequency

List of Subscripts

a – active, noncrystalline
c – cellulose, crystalline, conductivity
cw – cell wall substance
CD – cross direction, transverse direction
f – fiber saturation point, fiberboard, filler
k – final value
l – lignin
m – instantaneous, maximum value, most probable, mixture
MD – machine direction
r – relaxation, reactive
s – static
w – free moisture, wood
$\parallel$ – vector $\bar{E}$ is oriented parallel to the wood fibers or surface plane
$\perp$ – vector $\bar{E}$ is oriented perpendicular to the wood fibers or surface plane
0 – initial value
∞ – infinity

Chapter 1

Interaction between the Electromagnetic Field and Wood. Main Features of Dielectric Properties of Wood

Wood is a natural material with a complex structure and composition. Interaction between the alternating electromagnetic field and wood makes it possible to elucidate specific properties of this material. The electromagnetic field consists of two components: electric and magnetic fields. The influence of these components on wood is different. The influence of the magnetic field on wood is negligible and is not taken into consideration for practical purposes. The influence of the electric field on wood is very strong as the interaction between them results in the creation of electric currents in the material.

Under the action of an alternating electric field, the wood reveals its dielectric properties, which more often are characterized by two main indices: by the relative dielectric constant (ε') and by the dielectric loss tangent ($tg\delta$). The relative dielectric constant of a material shows how many times the force of interaction between the electric charges in the given medium is less than that in a vacuum. The relative dielectric constant of wood is equal to the ratio of the capacity of a condenser with the separator made of wood and the capacity of the vacuum capacitor of the same dimension and form.

The dielectric loss tangent of wood defines the part of the power applied to the wood that is absorbed by the material under the influence of the electric field. This part of the power is known to be transformed into thermal energy. The loss tangent is numerically equal to the ratio between the active current and the reactive current in the material or to the real and reactive powers ratio. One additional index is used in order to characterize the dielectric properties of wood and wood materials, viz. the dielectric loss factor $\varepsilon'' = \varepsilon' \cdot tg\delta$. The values of ε', ε'', and $tg\delta$ are always dimensionless, and their magnitude does not depend on the chosen system of units or the form of the equations.

1.1 Polarization of Wood

All materials are conventionally divided into three electric groups: conductors, dielectrics, and semiconductors. The conductors are characterized as materials with considerable electronic and ionic conductivity values. The dielectrics differ from the conductors by their low specific conductivity. Another princple distinction resides in the fact that the temperature coefficient of electric resistance of the dielectrics is negative while that of the conductors is positive. Concerning the specific conductance, the semiconductors occupy an intermediate position between the dielectrics and the conductors. By the character of the temperature dependence of conductivity, the semiconductors are close to the dielectrics.

It is assumed that conductors have a specific resistance in the range from about 10^{-8} to 10^{-5} ohm·m, semiconductors in the range from 10^{-6} to 10^{+9} ohm·m, and dielectrics in the range from 10^{+7} to 10^{+17} ohm·m (Tareev 1982). Oven-dry wood has specific resistances ranging from 10^{13} to 10^{15} ohm·m and is classified as a polar dielectric. As the wood moisture content increases, the specific resistance of wood becomes lower, and its conductivity approaches that of the semiconductors. When the moisture content exceeds the cell wall's saturation point, the wood may have an ionic conductance. Under the influence of an alternating electromagnetic field, the electric properties of wood are defined by polarization processes that take place because of the interaction between the molecules of the wood substance and the external field. In this case, moist wood as well as dry wood are considered polar dielectrics.

The magnetic properties of wood make it possible to ragard it as a diamagnetic material, i.e., as a substance that under the influence of an external magnetic field is magnetized in the direction opposite to the direction of the magnetic induction vector of the field. The magnetic properties of a substance are characterized by its magnetic susceptibility, which defines the relation between the vector of the magnetization of the material and the vector of the magnetic field intensity. The magnetic susceptibility of all diamagnetic materials (including wood and water) is a negative quantity. It depends neither on the temperature nor on the magnetic field strength. The magnetic susceptibility of wood has a very low absolute value (from -0.2×10^{-6} to -0.4×10^{-6}). Therefore, in the interaction between wood and an electromagnetic field, the influence of the magnetic component of the field is negligible and is not taken into account.

One of the most important intrinsic properties of wood is its polarization ability. The polarization effect is caused by the change in the arrangement in space of the electrically charged particles of the wood substance under the influence of an external electric field. Under this condition, wood acquires an electric moment.

In a theoretical examination of the properties of wood, an elementary volume is assumed to exist. This elementary volume should comprise as many structural units as are necessary for considering (with due accuracy) the assumed volume of wood as a homogeneous body. At the same time, the chosen volume should be sufficiently small so that the change in physical properties of the fields along its length could be neglected, and the properties can be considered uniform.

Sobolev (1979) proposed that substance elements of a size above 0.1 mm should be regarded as macrostructural elements of wood. According to this suggestion, to this category should be attributed such elements as layers of earlywood and latewood, rays of medium and large width, large resin canals, and large vessels. In this case, the assumption that the elementary volume should be not less than 1 cm^3 can be accepted. This volume should comprise a sufficient quantity of macroelements, and the indices of physical properties of the specimen should reflect not only the combined properties of the microelements but also the properties of the macroelements. When investigating the dielectric properties of wood, it is necessary to take into account the macrostructure of the wood, and therefore the samples size should be equal to or larger than the macrostructural elementary volume, i.e., $\geqslant 1\ cm^3$.

Polarization $\bar{P}$ is a quantity which numerically characterizes the substance polarization effect under the influence of the external electric field. In the absence of an electric field, each individual elementary volume of wood does not possess an electric moment as the algebraic sum of all charges of the molecules in the given volume is equal to zero.

The influence of the external electric field causes the ordering of the molecular charges in the wood substance. The volume of wood observed receives a certain electric moment equal to the geometric sun $\Sigma\bar{p}$ of the moments of all polarized molecules that constitute the given volume. The polarization intensity $\bar{P}$ is equal to the ratio limit of the electric moment of the observed volume of the material to the volume itself, when it tends to zero. In the most simple case, when the field is homogeneous, the polarization intensity $\bar{P}$ is the mean value of the summary dipole moment of the volume unit

$$\bar{P} = \Sigma\bar{p}/V\ . \tag{1.1}$$

The dielectric polarization intensity depends on the electric field strength $\bar{E}$. If this dependence is expressed through a linear differential equation, the dielectric is called linear. Wood is a linear dielectric for which

$$\bar{P} = \chi\varepsilon_0\bar{E}\ , \tag{1.2}$$

where

χ = electric susceptibility

ε_0 = electric constant which characterizes the electric field when its interaction with the substance is absent (in vacuum)

$$\varepsilon_0 = 8.854 \times 10^{-12}\ F/m\ .$$

The quantitative estimation of the substance polarizability is expressed through its relative dielectric constant ε'. The product $\varepsilon_0\varepsilon'$ (dimensions of quantities is F/m) is called the absolute dielectric constant of the substance.

The geometric sum of the electric field strength vector at a given point of the dielectric multiplied by the electric constant and of the polarization vector gives the electric displacement (electric induction) vector $\bar{D}$ at this point

$$\bar{D} = \varepsilon_0\bar{E} + \bar{P}\ . \tag{1.3}$$

The directions of vectors $\bar{P}$ and $\bar{E}$ in isotropic materials coincide. In anisotropic dielectrics, the electric susceptibility depends on the $\bar{E}$ direction and is a tensor of rank two.

The connection between the relative dielectric constant and the electric susceptibility χ is expressed in the form $\varepsilon' = 1 - \chi$ and the polarization of the material

$$\bar{P} = (\varepsilon' - 1)\varepsilon_0\bar{E}\ . \tag{1.4}$$

The dipole moment of the material volume unit can be represented (von Hippel 1954) as a result of additive actions of N elementary dipole moments $\bar{p}$, which are proportional to the local strength of the electric field $\bar{E}'$

$$\bar{P} = aN\bar{E}' , \tag{1.5}$$

where

a = proportionality factor, termed the polarizability ($F \cdot m^2$).

Equations (1.4) and (1.5) are two expressions for polarization connecting the macroscopic concept of the dielectric constant with the following molecular parameters: the polarizability a, the quantity of the molecular dipoles N, and the local strength of the electric field $\bar{E}'$. These parameters are required for the study of the dielectric polarization phenomena as a function of frequency, temperature, state of aggregation of the substance, and the external field strength. The total polarization of the substance, which represents the mean dipole moment of the unit of volume, is expressed as the sum of all types of polarization (Netushil et al. 1959). The summary polarization of wood, P, includes five kinds of polarization which take place in moist heterogeneous dielectrics

$$P = P_e + P_a + P_d + P_v + P_z , \tag{1.6}$$

where

P_e = electronic polarization
P_a = ionic (atomic) polarization
P_d = dipole (orientational) polarization
P_v = interfacial (structural) polarization
P_z = electrolytic polarization.

Electronic polarization arises as a result of the shift of the electron orbits relative to the positively charged nucleus under the influence of an external electric field. The electronic polarization takes place in all the atoms of any kind of substance. The electronic polarization sets in a very short time after the application of an electric field, viz. within $10^{-14} - 10^{-16}$ s. Because of this, it does not depend practically on the temperature and the field frequency, and it is not connected with the loss of energy.

Ionic (atomic) polarization arises as a result of an elastic displacement of atoms in the molecules as well as because of mutual displacement of charged ions of opposite signs in substances with ionic bonds. The centers of the positive and the negative charges of the cell ions that had coincided before the application of the electric field are driven some distance apart, and the cell acquires an electric moment. The polarization increases with a rise in temperature, as the thermal expansion drives ions away from each other and thus diminishes the forces interacting between them. The time it takes to produce ionic polarization is about 10^{-13}–10^{-12} s.

Dipole (orientational) polarization consists in rotation of dipole molecules in the direction of an external electric field. The effect of the dielectric polarization is created by the orientation of the dipole molecules, which previously had been in random thermal motion. The dipole polarization according to its nature is closely linked with the thermal motion of the molecules, and therefore this kind of polarization is influenced by the temperature. Dipole polarization manifests itself in gases, liquids (water), and amorphous substances. It is also present in some crystal bodies, e.g., in ice. Rotation may also occur in the polymers or a change of state and not only of the whole molecules but also of their parts as well. The macromolecules cellulose, hemicellulose and lignin, which make up the cell wall substance in wood, are also subject to dipole polarization. Such a process is conditioned by the displacement of polar groups such as OH or CH_2OH relative to the motionless parts of the macromolecules. This type of polarization is termed the *dipole-radical polarization*. The process of a gradual increase or decrease of the polarization of a material, which follows an increase or decrease of the electric field strength, is called the dielectric relaxation. After removal of the field, the dipole polarization is distorted by the thermal motion of the molecules, and its value diminishes in accordance with the exponential law

$$P_d(\tau) = P_d(0)\exp(-\tau/\tau_0)\,, \tag{1.7}$$

where

$P_d(0)$ = polarization level at the moment of the field removal
τ_0 = time constant of the polarization decrease process (the dipole polarization relaxation time)
τ = current time.

The relaxation time is the period of time that is required for the decrease of the dielectric polarization level by $e = 2.72$ times, relative to the initial value, after the removal of the field. Usually τ_0 ranges from 10^{-6} to 10^{-12} s, and this is the reason why the dipole polarization manifests itself at frequencies below 10^{12} Hz. The value of τ_0 increases considerably with a decrease in temperature. The dipole polarization causes a dissipation of the electric energy, which in wood is transformed into heat.

In the nonuniform (heterogeneous) dielectrics, which are composed of substances with different properties, the so-called *interfacial polarization* takes place. The moist wood is a heterogeneous material composed of substances in solid, liquid, and gaseous phases. The solid phase consists of the wood cell walls, the surfaces of which absorb moisture, forming bound water. Inside the cell walls, the interfibril channels and cell wall cavities are partially filled with air and vapor. If the quantity of moisture exceeds the fiber saturation point, the cell cavities become filled with water.

Because of the contact difference of potentials between the water and the cell wall substance, the molecules of water receive a charge of one sign, while the cell wall substance assumes a charge of opposite sign. The cell wall substance has considerably less electric conductivity than the water. On application of an electric field, free electrons and ions, which are present in the conductive and semiconductive elements in the cell walls, start moving within each element's volume, and as a result the element receives a dipole moment. Figure 1.1b

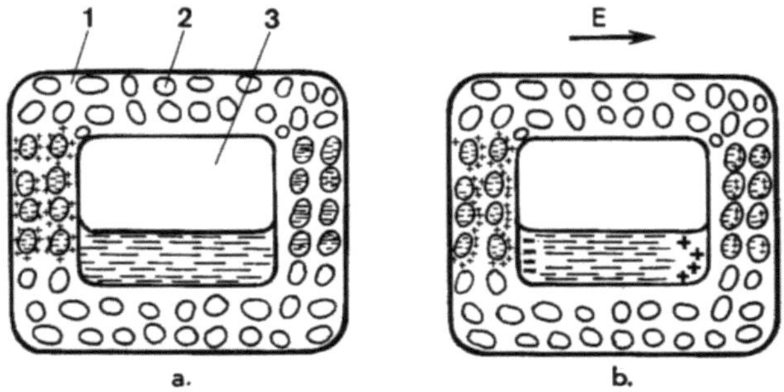

Fig. 1.1a, b. Diagrammatic representation of wood cell cross-section. **a** External electric field is absent; **b** polarization after the imposition of an external electric field *E*. *1* The cell wall; *2* microcapillary tubes; *3* the cell cavity

schematically shows the result of the polarization of the conductive elements in wood caused by the conduction current. The structural heterogeneity of the wood results in a limited displacement of the charges, which, being free within each cavity or channel, behave themselves in the same manner as bound charges. The dissociated ions that are present in the free water of the cell cavity shift their position under the influence of the imposed field and thus polarize the water. At the instant of imposition of the electric field, the mean dipole moment of the substance volume unit, which characterizes the interfacial polarization P_v, increases gradually, a process that is determined by the relaxation time. For the ionic kind of element conductivity, which usually takes place in moist wood, the time constant of the interfacial polarization varies from 10^{-3} to 10^{-8} s.

In moist materials, the application of DC voltage causes an electrolysis process due to the displacement of dissociated ions. The material always has a residual electrochemical or electrolytic polarization. At the instant of imposition of an electric field, the increase of the *electrolytic polarization* is observed. The increase of this kind of polarization is much slower than that of the other types of polarization considered above. The electrolytic polarization of wood can be characterized by its time constant, which is approximately equal to 10^{-4}–10^{2} s.

Under the action of an electric field on wood, all the described kinds of polarization take place. However, depending on the field frequency, the effect exerted by each kind of polarization on the general polarization process is different. This means that their influence on the dielectrical properties of wood also varies.

The general polarization of wood, *P*, can be regarded as a total of two components, namely instant polarization (P_m) and relaxation polarization (P_r), so that $P = P_m + P_r$. The instant polarization is the sum of the electronic and ionic (atomic) polarization. These kinds of polarization follow instantly electric field strength changes. The total of dipole, interfacial, and electrolytic polarizations of wood constitutes the relaxation component of the general polarization.

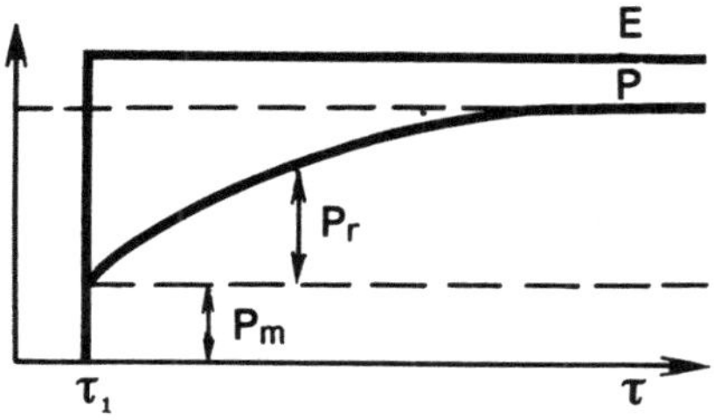

Fig. 1.2. Diagram of the increase in wood polarization P upon an increase in the electric field strength E: P_m, P_r the instant and the relaxation components of polarization; τ, τ_1 time

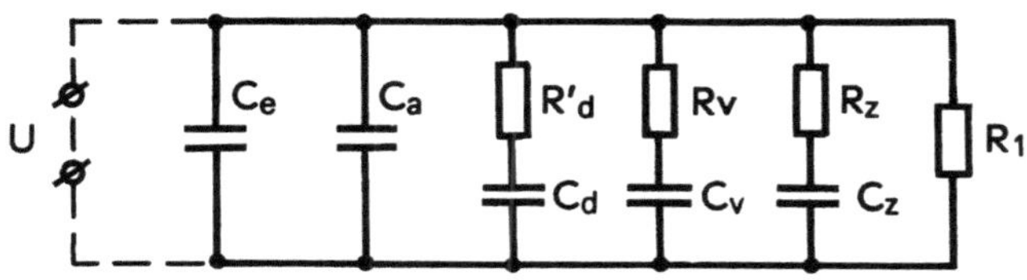

Fig. 1.3. Equivalent circuit of wood as a dielectric with different kinds of polarization: C_e, C_a, C_d, C_v, C_z capacitance values of the condensers which correspond to electronic, ionic (atomic), dipole, interfacial, and electrolytic types of polarization; R_d, R_v, R_z resistors equivalent to the energy losses at dipole, interfacial, and electrolytic kinds of polarization; R_l resistance of the dielectric to the conductivity of the direct current

This component lags behind the changes of the field strength. Figure 1.2 shows the increase of the general polarization of wood after an instant increase of the field strength E.

The equivalent circuit of wood as a dielectric material in which different polarization mechanisms exist can be represented by a circuit comprising capacitors C and resistors R (Fig. 1.3). Resistor R_1 represents the resistance of wood to the direct current conduction.

Not all kinds of polarization contribute equally to the general polarization of wood. This is the reason why under certain conditions, depending on the frequency range, some kinds of polarization are not taken into consideration. Thus, in the range 10^5–10^{10} Hz, which are used for high frequency (HF) and UHF heating, the influence of electrolytic polarization (a time constant of about 10^{-4}–10^2 s) and of electronic and ionic polarizations (10^{-12}–10^{-16} s) is inconspicuous. Dipole and interfacial kinds of polarization play the main role in the polarization of wood.

The cell wall substance consists of molecules of different substances. Wood also contains moisture in various forms and states. Molecules of these substances and of water have different relaxation times. This is the reason why the distribution of the relaxation time τ and of wood polarizability is subject to statistical laws that can be defined by a certain normalized distribution function

$$\int_0^\infty G(\tau)d\tau = 1 \,. \tag{1.8}$$

The influence of an external electric field that changes with time excites an alternating electric current in the wood. The general density of this current is usually represented by a complex value J^* equal to the sum of the conduction current, J_c, and the displacement current, J_p

$$J^* = J_c + J_p . \tag{1.9}$$

The conduction current

$$J_c = \sigma \bar{E} , \tag{1.10}$$

where

σ = specific conductance not depending on $\bar{E}$.

The displacement current consists of two components: (a) the instant displacement current

$$J_m = \varepsilon_0 \cdot \partial \bar{E}/\partial \tau + \partial \bar{P}/\partial \tau , \tag{1.11}$$

where

$\varepsilon_0 \partial E/\partial \tau$ = the displacement current density, which is conditioned by the electric field changes vs. time

$\partial \bar{P}/\partial \tau$ = the current density determined by the electronic and ionic (instant) polarization;

(b) the relaxation displacement current

$$J_r = \partial \bar{P}_r/\partial \tau , \tag{1.12}$$

which is determined by other kinds of polarization.

The complex current density can be defined as a sum of the three components

$$J^* = J_c + J_m + J_r . \tag{1.13}$$

The conduction current density is in phase with the electric field strength (Fig. 1.4). In the process of electronic and ionic polarization, the charged

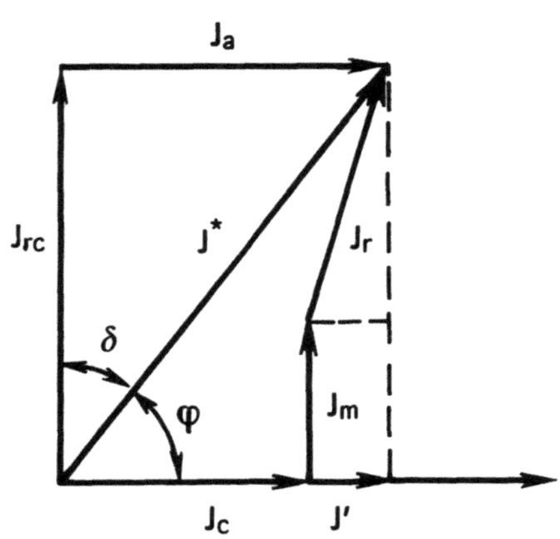

Fig. 1.4. Diagram of currents in wood under the action of an external alternating current field E: J^* total current; J_a active component of the current; J_{rc} reactive component of the current; J_c conduction current; J_m instant displacement current; J_r relaxation displacement current; J' active component of the displacement current; δ dielectric loss angle; φ phase displacement angle

particles of the dielectric instantly follow the field changes, and the phase of the polarization vector coincides with that of the field strength. The phase lead of the instant displacement current relative to the electric field strength is $\pi/2$ radians.

In the dipole and interfacial polarization processes, the displacement of dielectric particles lags behind the electric field strength, i.e., the relaxation process takes place. In this case, the phase of the polarization vector does not coincide with the field strength vector, and the displacement current density forestalls it by a little less than $\pi/2$ radians. A component J' emerges in the displacement current density. This component is in phase with the field strength, its appearance is caused by the fact that a certain quantity of energy should be spent to surmount frictional forces that oppose the polarization process. This energy is released in the dielectric material in the form of heat.

The total conduction current is called the active component, J_a, and the displacement current is termed the reactive component, J_{rc}, of the summary current in the dielectric. The total current density J^* forms an angle with the electric field strength E equal to $\varphi = \pi/2 - \delta$.

The angle δ is called the dielectric loss angle. The tangent of this angle is equal to the ratio of the active and reactive currents or to the ratio of the real power N_a and the reactive power N_r

$$tg\delta = J_a/J_{rc} = N_a/N_r \,. \tag{1.14}$$

1.2 Dielectric Parameters of Wood

To facilitate the determination of the power that is released in a dielectric and the current density in the dielectric, the physics of dielectric materials introduced such a concept as the complex dielectric constant ε^*, which entails examination of both the conduction and the displacement currents. Using ε^* and knowing the electric field strength E, it is easy to determine the active and the reactive components of the current density and also the real and the reactive specific powers.

The dielectric properties of wood can also be determined through the complex dielectric constant

$$\varepsilon^* = \varepsilon' - i\varepsilon'' = \varepsilon'(1 - itg\delta) \,, \tag{1.15}$$

where

ε' = the relative dielectric constant (real part)
ε'' = the loss factor (imaginary part)
$tg\delta$ = loss tangent, $tg\delta = \varepsilon''/\varepsilon'$
i = $\sqrt{-1}$.

Such components of Eq. (1.15) as the dielectric constant ε' and $tg\delta$ are the principal dielectric parameters of wood, and they shall be so considered below.

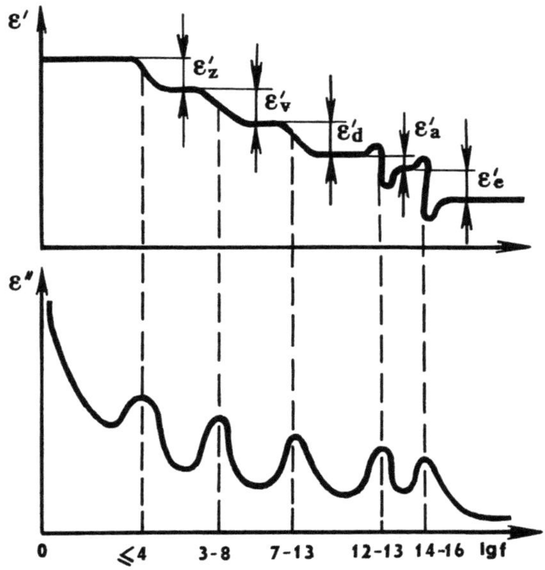

Fig. 1.5. Schematic diagram of dielectric constant ε' and loss factor ε'' of wood versus frequency response characteristics at different kinds of polarization. The contribution of different kinds of polarization into ε' value: ε'_e electronic polarization deposit; ε'_a ionic (atomic) polarization deposit; ε'_d dipole polarization deposit; ε'_v interfacial polarization deposit; ε'_z electrolytic polarization deposit; *lg f* logarithm of frequency value

In some cases, it is convenient to use the loss factor $\varepsilon'' = \varepsilon' \times tg\delta$. The dependence of the dielectric constant on the frequency $\varepsilon' = \varphi(f)$ is called the dielectric constant dispersion of wood. The influence of different kinds of polarization on the wood dielectric constant ε' and on the loss factor ε'' is shown in Fig. 1.5.

With an increase in frequency, the relaxation polarization leads to a monotonic decrease in ε', which can be explained by the gradual decrease in the quantity of dipoles and charges that carry out their displacement or a change in orientation following the changes in the external field. The maximum value of the loss factor ε'' always corresponds to the relaxation and resonance frequencies. The increase in the loss factor at low frequencies may be explained by the increase in the direct current conductance. At frequencies of 10^{12}–10^{16} Hz, a resonance polarization of electronic and ionic (atomic) displacement character is observed in the dielectrics. When approaching the resonance frequency, ε' begins to increase, as at frequencies which are near the natural oscillation frequency of the electron orbit or of ions, their atomic vibrations increase. Wood may have several resonance frequencies as it consists of several different substances. At frequencies somewhat higher than the resonance frequency, the polarization component proves to be in antiphase to the external field, and the dielectric constant ε' diminishes.

If the relaxation time τ of dipoles of the dielectric is the same, then the connection between the complex dielectric constant and the frequency is to be found from the Debye equation

$$\varepsilon^* = \varepsilon'_\infty + (\varepsilon'_s - \varepsilon'_\infty)/(1 + i\omega\tau) , \tag{1.16}$$

where

ε'_∞ = the limiting high-frequency relative dielectric constant
ε'_s = the limiting low-frequency relative dielectric constant (static)
ω = the angular frequency
τ = the macroscopic time of relaxation.

In deriving Eq. (1.16) it was supposed that the polarization process depends on the time τ_1 as follows:

$$P = P_0 \exp(-\tau_1/\tau) . \tag{1.17}$$

Extracting the real and imaginary parts from Eq. (1.16) we obtain the frequency response of ε' and ε''

$$\varepsilon' = \varepsilon'_\infty + (\varepsilon'_s - \varepsilon'_\infty)/(1 + \omega^2\tau^2) , \tag{1.18}$$

$$\varepsilon'' = (\varepsilon'_s - \varepsilon'_\infty)\omega\tau/(1 + \omega^2\tau^2) . \tag{1.19}$$

The peak of ε'' occurs at $\omega_m\tau = 1$, where

ω_m = the angular frequency corresponding to the maximum absorption value.

The wavelength corresponding to the maximum absorption value is derived from

$$\lambda_m = 2\pi c/\omega_m , \tag{1.20}$$

where

c = the velocity of light.

For determination of the frequency dependence of the dielectric constant and the loss factor of polar substances, Cole and Cole (1941) suggested a diagram in which the loss factor is plotted on the ordinate and the dielectric constant on the abscessa, while the angular frequency ω is a variable quantity. These authors proved by experiments that if ε'' as a function of ε' of a substance with one time of relaxation is plotted in this coordinate system, a semicircle is obtained with the semicircle diameter positioned on the ε' axis. The intersection points of the semicircle with the abscissa give values of static ε'_s and limiting high-frequency ε'_∞ of the dielectric constant (Fig. 1.6a). The elimination of $\omega\tau$ from Eqs. (1.18) and (1.19) results in the equation of a circumference in the coordinate system of (ε', ε''), as follows:

$$[\varepsilon' - (\varepsilon'_s + \varepsilon'_\infty)/2]^2 + (\varepsilon'')^2 = [(\varepsilon'_s - \varepsilon'_\infty)/2]^2 . \tag{1.21}$$

The radius of the circumference is equal to $(\varepsilon'_s - \varepsilon'_\infty)/2$, and the central point of the circumference is on the x-axis at a distance of $(\varepsilon'_s + \varepsilon'_\infty)/2$ from the origin of the coordinates (Fig. 1.6a).

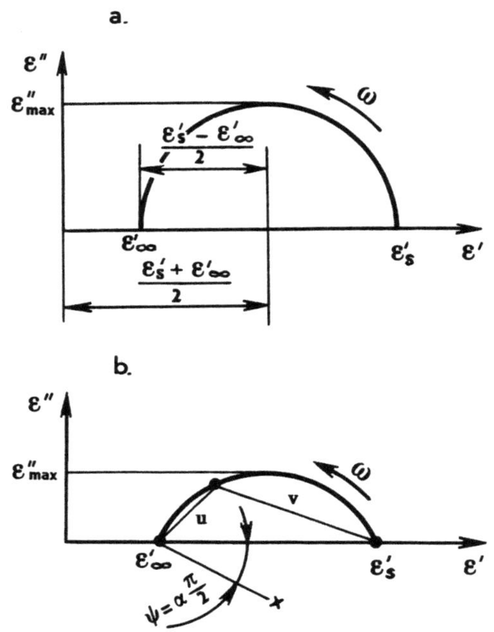

Fig. 1.6a, b. Cole-Cole diagrams: *a* for a material with a single relaxation time; *b* for a material with a distributed relaxation time. α Parameter of distribution of the relaxation time; ε' dielectric constant; ε_s' limiting low-frequency dielectric constant (static); ε_∞' limiting high-frequency dielectric constant (optic); ε'' loss factor; ε''_{max} maximum value of the loss factor; ω angular frequency; ψ angle; U, V distances from an experimental point to the points ε_∞' and ε_s'

The applicability of the Debye equations can be verified by plotting ε'' as a function of ε' with the use of empirical data obtained at different frequencies. If the Debye equations can be used for a given material, this dependence would be described by a semicircle, the intersections of which with the abscissa on the right-hand side when $\omega = 0$ gives the value of ε_s' and on the left-hand side when $\omega \to \infty$, the value of ε_∞'. The maximum of the dependence of ε'' on ε' is at $\omega\tau = 1$. Measurements show that the Debye equations can be applied to polar liquids whose molecules are nearly spherical. Within the experimental error, the Debye equations describe the dielectric parameters of water with a good degree of approximation.

If the dielectric consists of different molecules, and each molecule has its own polarization setting time, a straggling of the relaxation times takes place, so that they are distributed around the most probable value.

For cases when the distribution of the relaxation time occurs, Cole and Cole (1941) suggested the following empirical equation instead of Eq. (1.16):

$$\varepsilon^* = \varepsilon_\infty' + (\varepsilon_s' - \varepsilon_\infty')/[1 + (i\omega\tau)^{1-\alpha}] , \tag{1.22}$$

where

α = the parameter which characterizes the relaxation time distribution; it varies from 0 to 1.

When $\alpha = 0$, this equation is reduced to the Debye equation. Separation of the real and the imaginary parts in Eq. (1.22) at $\alpha > 0$ and exclusion of $\omega\tau$ reduce the equation to that of a circumference with the center point below the abscissa (Fig. 1.6b). The angle ψ formed by the x-axis and the line connecting the circumference center point with the point $\varepsilon' = \varepsilon'_\infty$ makes it possible to determine the relaxation time distribution parameter α using the following equation:

$$\psi = \alpha \frac{\pi}{2} \,. \tag{1.23}$$

The relaxation time can be derived from the relation:

$$V/U = (\omega\tau)^{1-\alpha} \,, \tag{1.24}$$

where

V and U are the distances from an experimental point on the semicircle to the points $\varepsilon' = \varepsilon'_s$ and $\varepsilon' = \varepsilon'_\infty$, respectively

$$V = \sqrt{(\varepsilon'_s - \varepsilon')^2 + (\varepsilon'')^2} \,; \quad U = \sqrt{(\varepsilon' - \varepsilon'_\infty)^2 + (\varepsilon'')^2} \,. \tag{1.25}$$

The molecules of the substances forming wood possess different relaxation times. This is the reason why the Eqs. (1.22) through (1.25) are used for the analysis of the dielectric properties of wood.

The dielectric constant and loss tangent are connected with the active σ' and the reactive σ'' components of the complex specific conductance through the following equations:

$$\sigma' = \omega\varepsilon'\varepsilon_0 tg\delta \quad (\text{ohm}^{-1}\cdot\text{m}^{-1}) \,, \tag{1.26}$$

$$\sigma'' = \omega\varepsilon'\varepsilon_0 \quad (\text{ohm}^{-1}\cdot\text{m}^{-1}) \,, \tag{1.27}$$

where

ω = the angular frequency.

Parameters ε' and $tg\delta$ participate in all calculations connected with the interactions between an electromagnetic field and wood and depend on the tree species, wood density, water content, and temperature, as well as on the field frequency and its orientation in relation to the direction of the grain. Temperature and frequency act upon the dielectric properties of wood as factors of an external medium.

The dielectric properties of wood are distinguished in the three directions that are generally adopted for examinations of wood structure, namely, the longitudinal L, radial R and tangential T directions (Fig. 1.7). If the properties of wood are described in relation to a direction (e.g., in the radial direction) it means that vector $\bar{E}$ is oriented in the same (radial) direction, i.e., at low and high frequencies the electrodes imposing a voltage should be appliled to the tangential sides of the sample. If the dielectric properties of wood in the longitudinal direction are being considered, this means that vector $\bar{E}$ is parallel to the longitudinal axis of the stem, i.e., at low and high frequencies the electrodes are to be applied to the end-face planes of the specimens. If the electric field vector $\bar{E}$ is parallel with the longitudinal axis of the stem (L), the dielectric constant and

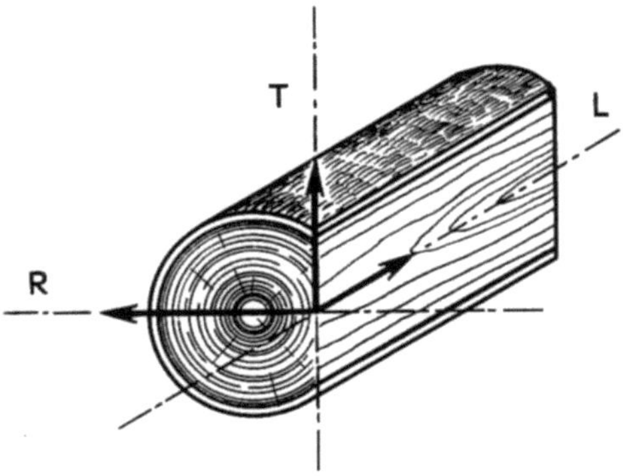

Fig. 1.7. Principal axes of wood anisotropy: *L* longitudinal axis; *R* radial axis; *T* tangential axis

loss tangent should be written as ε'_L and $tg\delta_L$ or $\varepsilon'_{||}$ and $tg\delta_{||}$. If vector $\bar{E}$ is parallel with the radial or tangential directions, these parameters should be designated as ε'_R, $tg\delta_R$ and ε'_T, $tg\delta_T$, respectively. If the vector $\bar{E}$ is oriented in a transverse direction to the *L* direction without any indication of its direction in relation to the *R* and *T* axes, the designations of ε' and $tg\delta$ shall be $\varepsilon'_{\perp}$ and $tg\delta_{\perp}$, respectively. This is acceptable since in many cases the difference between the properties of wood in the radial and tangential directions can be neglected. When the mean values of the radial and tangential, radial and longitudinal, and tangential and longitudinal parameters of wood are being considered, the following designations are used: ε'_{RT}, ε'_{RL}, ε'_{TL}, $tg\delta_{RT}$, $tg\delta_{RL}$, $tg\delta_{TL}$.

In studying the processes of interaction between an electric field and wood in those cases in which the wavelength considerably exceeds the cell dimensions, it becomes possible to regard wood as a continuous medium. If the wavelength is less than 1 cm, its dimensions can be compared with those of the cells, and the influence of separate cells and of their components on the interaction process become significant.

The anisotropy of wood is a result of its gross anatomy and of the microscopic structure of its cells. The microfibrils in the wood cell walls are oriented relative to the longitudinal axis of the cell. However, the character of the anisotropy of these elements cannot be applied to all wood elements. The anatomy of wood is the most important factor influencing the anisotropy.

Wood is generally regarded as an orthogonal isotropic material, i.e., it is a material that has three mutually perpendicular structural symmetry planes. This is an idealized scheme as wood properties vary considerably along the radius and the height of a stem, and also in relation to the moisture content. In addition, wood is of an irregular constitution, which is characteristic of each separate tree, etc. However, wood is generally considered as a material whose elementary volumes have properties of orthogonal anisotropy.

The symmetry axes of an orthotropic material are called the principal axes of anisotropy. The principal axes of wood are the axes directed along the tree stem (*L*), in the radial (*R*), and in the tangential (*T*) directions.

The dielectric properties of wood with its intrinsic anisotropy taken into consideration can be described by means of the second rank tensor

$$(\varepsilon') = \begin{vmatrix} \varepsilon'_{LL} & \varepsilon'_{LR} & \varepsilon'_{LT} \\ \varepsilon'_{RL} & \varepsilon'_{RR} & \varepsilon'_{RT} \\ \varepsilon'_{TL} & \varepsilon'_{TR} & \varepsilon'_{TT} \end{vmatrix} \tag{1.28}$$

and

$$(tg\delta) = \begin{vmatrix} tg\delta_{LL} & tg\delta_{LR} & tg\delta_{LT} \\ tg\delta_{RL} & tg\delta_{RR} & tg\delta_{RT} \\ tg\delta_{TL} & tg\delta_{TR} & tg\delta_{TT} \end{vmatrix} \cdot \tag{1.29}$$

It is generally accepted that the dielectric properties of wood do not change with the revolution of $\bar{E}$ through 180° and that the axes L, R, and T are the principal axes of the tensor. Therefore, Eqs. (1.28) and (1.29) are simplified to:

$$(\varepsilon') = \begin{vmatrix} \varepsilon'_{L} & 0 & 0 \\ 0 & \varepsilon'_{R} & 0 \\ 0 & 0 & \varepsilon'_{T} \end{vmatrix} \tag{1.30}$$

and

$$(tg\delta) = \begin{vmatrix} tg\delta_{L} & 0 & 0 \\ 0 & tg\delta_{R} & 0 \\ 0 & 0 & tg\delta_{T} \end{vmatrix} \cdot \tag{1.31}$$

If the values ε' and $tg\delta$ are computed at $\bar{E}$ oriented parallel to the principal axes of symmetry, the values of ε'_x and $tg\delta_x$ in the arbitrary direction of $\bar{E}$ forming angles α_1, α_2, and α_3 with the axes L, R, and T, respectively, are determined from the following equations:

$$\varepsilon'_x = \varepsilon'_L \cos^2\alpha_1 + \varepsilon'_R \cos^2\alpha_2 + \varepsilon'_T \cos^2\alpha_3 \ , \tag{1.32}$$

$$tg\delta_x = tg\delta_L \cos^2\alpha_1 + tg\delta_R \cos^2\alpha_2 + tg\delta_T \cos^2\alpha_3 \ . \tag{1.33}$$

The geometric interpretation of the dielectric parameters of wood, ε' and $tg\delta$, is an ellipsoid (Fig. 1.8) whose axes are directed along axes L, R, and T. The lengths of the radius vectors, which are directed along the ellipsoid axes, are as follows: $1/\sqrt{\varepsilon'_L}$, $1/\sqrt{\varepsilon'_R}$, $1/\sqrt{\varepsilon'_T}$ or $1/\sqrt{tg\delta_L}$, $1/\sqrt{tg\delta_R}$, $1/\sqrt{tg\delta_T}$. The length r of any radius vector of the characteristic surface of the ellipsoid is numerically equal to the square root of the quantity $1/\varepsilon'(1/tg\delta)$, which characterizes the dielectric constant ε' (or $tg\delta$) in the same direction.

For the purpose of analysis of experimental data, it is convenient to characterize the relations of the dielectric parameters of wood in the directions of the

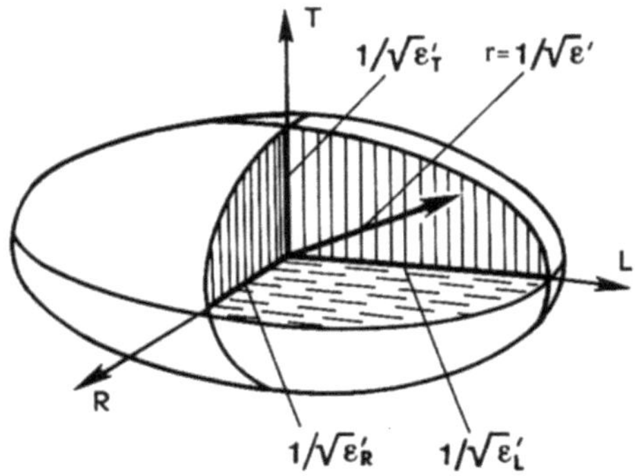

Fig. 1.8. Characteristic ellipsoid of the wood dielectric constant: *L*, *R*, *T* longitudinal, radial, and tangential axes of anisotropy; ε' dielectric constant at the given orientation of the electric field strength vector; ε'_L, ε'_R, ε'_T dielectric constant values in the longitudinal, radial, and tangential directions; *r* radius vector of the characteristic surface of the ellipsoid

principal axes of symmetry with coefficients:

$$k_1 = \varepsilon'_{||}/\varepsilon'_{\perp}$$
$$k_2 = tg\delta_{||}/tg\delta_{\perp} \tag{1.34}$$
$$k_3 = \varepsilon''_{||}/\varepsilon''_{\perp} ,$$

$$k_{1(R/T)} = \varepsilon'_R/\varepsilon'_T$$
$$k_{2(R/T)} = tg\delta_R/tg\delta_T \tag{1.35}$$
$$k_{3(R/T)} = \varepsilon''_R/\varepsilon''_T .$$

If the difference of wood parameters in the radial and tangential directions is negligible (which often happens), the values of $\varepsilon'_{||}$ and $tg\delta_{||}$ can be expressed through coefficients k_1 and k_2 as follows:

$$\varepsilon'_{||} = k_1 \varepsilon'_{\perp}$$
$$tg\delta_{||} = k_2 tg\delta_{\perp} .$$

Then, Eqs. (1.32) and (1.33) can be simplified:

$$\varepsilon'_x = \varepsilon'_{\perp}(k_1 \cos^2\alpha_1 + \sin^2\alpha_1) , \tag{1.36}$$
$$tg\delta_x = tg\delta(k_2 \cos^2\alpha_1 + \sin^2\alpha_1) , \tag{1.37}$$

where

α_1 = angle between the direction of $\bar{E}$ and axis *L*.

The influence of the temperature on the dielectric properties of wood is essential. The estimation and calculation of this influence can be accomplished through temperature coefficients (Tareev 1982):

$$k^t_{\varepsilon'} = (1/\varepsilon')\cdot(d\varepsilon'/dT)(\mathrm{K}^{-1}) , \tag{1.38}$$

$$k^t_{tg\delta} = (1/tg\delta)\cdot(dtg\delta/dT)(\mathrm{K}^{-1}) , \tag{1.39}$$

where

T = the absolute temperature, K.

If the dependence of ε' and $tg\delta$ upon the temperature is linear, their value at any given temperature can be derived from the following equations:

$$\varepsilon' = \varepsilon'_{20}[1 + k^t_{\varepsilon'}(T - 293)] , \tag{1.40}$$

$$tg\delta = tg\delta_{20}[1 + k^t_{tg\delta}(T - 293)] , \tag{1.41}$$

where

ε'_{20} and $tg\delta_{20}$ are the dielectric constant and dielectric loss tangent at +20 °C.

1.3 Some Principal Formulas Describing Interactions Between an Electromagnetic Field and a Material

The practical application of the results of interactions between electromagnetic fields and wood and wood-based materials is carried out in two main directions: (1) in technological processes of power engineering and (2) in measuring devices used for the inspection of the parameters of various materials.

For the purposes referred to in item 1, powerful sources of electromagnetic radiation (up to thousands of kilowatts) are required, while for the devices in item 2, only low-powered (estimated in microwatts) sources are used.

In all calculations connected with the processing of wood and wood-based materials as well as with the measurement of various properties of these materials, such dielectric parameters of wood as ε', $tg\delta$, and ε'' are widely used. These parameters depend upon the moisture, the density, the structure, and the temperature of the materials as well as on the frequency and the orientation of the electric field strength vector.

The laws under which electromagnetic waves interact with dielectric materials have been thoroughly studied, and the results of this research are found in the literature (von Hippel 1954, 1959; Netushil et al. 1959). Below are given some basic formulas that characterize interactions between an electric field and dielectrics and are most often used in practical calculations.

The capacitance of a plane capacitor depends on the dielectric properties of the material that fills the space between the plates of the capacitor. This capacitance can be determined by the following equation:

$$C = \varepsilon'\varepsilon_0 Az/d , \tag{1.42}$$

where

A = the area of the capacitor plates

d = the distance between the plates
z = the number of spacings between the capacitor plates.

The force F of the interaction between the capacitor plates is defined as

$$F = \varepsilon' \varepsilon_0 E^2 A/2 , \tag{1.43}$$

where
E = electric field strength.

Energy accumulated by the capacitor is

$$W' = \varepsilon' \varepsilon_0 E^2 V_c/2 , \tag{1.44}$$

where
V_c = the volume of the capacitor.

When an electromagnetic wave acts on the dielectric, a part of the incoming energy is reflected, and the other part passes through the surface. This portion is propagated inside the dielectric and is absorbed by the material. If a plane electromagnetic wave acts perpendicular to the surface of the material, the reflection factor R_N, which characterizes the ratio of reflected to incident power, can be defined from the following equation:

$$R_N = (\sqrt{\varepsilon'} - 1)^2/(\sqrt{\varepsilon'} + 1)^2 . \tag{1.45}$$

The velocity v of electromagnetic wave propagation in nonmagnetic materials (to which wood and wood-based materials are related) depends on their properties and can be defined from:

$$v = c/\sqrt{\varepsilon'} , \tag{1.46}$$

where
c = velocity of light.

The wavelength in a material is

$$\lambda = \lambda_0/\sqrt{\varepsilon'} , \tag{1.47}$$

where
λ_0 = wavelength in vacuum (in the air).

The electromagnetic field penetrates into the material to a considerable depth. The electric field strength, equal to E_0 on the surface of the material, will be attenuated inside the material according to the exponential equation

$$E = E_0 \exp(-\alpha X) , \tag{1.48}$$

where
α = attenuation constant
X = distance from the surface of the dielectric.

The attenuation constant can be derived from the following equation:

$$\alpha = (2\pi/\lambda_0)[(\varepsilon'/2)\cdot(\sqrt{1 + tg^2\delta} - 1)]^{1/2} . \tag{1.49}$$

The attenuation value is often expressed through the attenuation distance Δ, beyond which the electromagnetic field strength is diminished by e times ($1/e = 0.368$) from the initial value

$$\Delta = 1/\alpha = (\lambda_0/2\pi)\cdot[2/\varepsilon'(\sqrt{1 + tg^2\delta} - 1)]^{1/2} . \quad (1.50)$$

The character of the dissipation of energy in wood and of its transformation into heat is determined by the values of the dielectric constant and the loss tangent. Propagation of the energy W' in the material to the depth X can be determined by the following equation:

$$W' = W'_0 \exp(-2\alpha X) , \quad (1.51)$$

where

W'_0 = energy applied to the surface layer of the material
α = attenuation constant derived from eq. (1.49).

Specific dielectric losses, i.e., the power emitted per unit volume of wood (W/m^3) under the influence of the electromagnetic field, are expressed by the following equation:

$$N = \omega\varepsilon'\varepsilon_0 tg\delta \cdot E^2 = 5.56 \times 10^{-11} f\varepsilon' tg\delta E^2 , \quad (1.52)$$

where

f = frequency in Hz
E = electromagnetic field strength in V/m.

The rate of heating of the material under the action of an electromagnetic field is determined through the following formula:

$$dt/d\tau = N/\rho c_p \quad (1.53)$$

or

$$dt/d\tau = 5.56 \times 10^{-11} f\varepsilon' tg\delta E^2/\rho c_p , \quad (1.54)$$

where

ρ = density of the material, kg/m^3
c_p = the heat capacitance of the material, $J/kg \cdot K$.

Chapter 2

Wood Composition and Dielectric Properties of Its Components

2.1 Physical Model of Wood as a Multicomponent Dielectric

A theoretical definition of the influence of the structure of wood on its dielectric properties is an extremely complicated task. At present, the theories of calculations of dielectric parameters of pure substances consisting of identical molecules and of two-component mixtures of the elementary substances are thoroughly worked out, e.g., the theories of Debye, Onsager, Klauzius-Mosotti, and Kirkwood-Frölich. Wood, however, consists of many different elements and substances. Today, three levels of nonuniformity of wood constitution and composition are distinguished, namely macrostructural (earlywood and latewood, rays, resin canals, and vessels), microstructural (cells, walls of cells and vessels, layers of cell walls), and molecular (molecules of cellulose, hemicelluloses, lignin, extractives, salts, water, and other substances). Practically speaking, it is impossible to take into account the influence of all levels of wood nonuniformity on its dielectric parameters. This is the reason why for a theoretical analysis of the dielectric properties of wood this consideration must be simplified to the assumption that wood is a physical composition (mixture) of only a few components. Thus, the analysis is carried out on the macroscopic level, assuming that the layers of the components contain a very large number of molecules that are uniform and isotropic, and that they possess uniform parameters through the whole volume of the layer. In this manner, certain parameters of a fictious uniform dielectric material can be determined. This material, when inserted between the plates of a capacitor, should create the same capacitance value as a capacitor of the same dimensions but with the plates separated by the heterogeneous dielectric material under exploration, namely, wood.

Oven-dry wood can be considered as consisting of two components: cell wall substance and air. In moist wood with the moisture level below the fiber saturation point, a third component must be added. This component is the bound water. At a moisture level exceeding the saturation point, a fourth component must be added, namely, free water. At negative temperatures, the place of the free water and of a part of the bound water is filled by ice. All the above-mentioned components of wood influence its dielectric parameters, and moist wood should accordingly be considered as a mixture of these components. Therefore, the physical model of an elementary section of moist wood should include all four components.

For the analytical expression of the dielectric parameters of such a composition use can be made of equations suggested for dielectric mixtures by Lichtenecker (1926) and Rother (for ε'_m), and Büchner (1939) (for $tg\delta_m$) (see Trapp and

Pungs 1956a):

$$(\varepsilon'_m)^k = V_1(\varepsilon'_1)^k + V_2(\varepsilon'_2)^k + V_3(\varepsilon'_3)^k + V_4(\varepsilon'_4)^k , \tag{2.1}$$

$$tg\delta_m \cdot k = \frac{V_1\left(\frac{\varepsilon'_1}{\cos\delta_1}\right)^k \cdot \sin\delta_1 \cdot k + V_2\left(\frac{\varepsilon'_2}{\cos\delta_2}\right)^k \cdot \sin\delta_2 \cdot k + V_3\left(\frac{\varepsilon'_3}{\cos\delta_3}\right)^k \cdot \sin\delta_3 \cdot k + V_4\left(\frac{\varepsilon'_4}{\cos\delta_4}\right)^k \cdot \sin\delta_4 \cdot k}{V_1\left(\frac{\varepsilon'_1}{\cos\delta_1}\right)^k \cdot \cos\delta_1 \cdot k + V_2\left(\frac{\varepsilon'_2}{\cos\delta_2}\right)^k \cdot \cos\delta_2 \cdot k + V_3\left(\frac{\varepsilon'_3}{\cos\delta_3}\right)^k \cdot \cos\delta_3 \cdot k + V_4\left(\frac{\varepsilon'_4}{\cos\delta_4}\right)^k \cdot \cos\delta_4 \cdot k}, \tag{2.2}$$

where

$\varepsilon'_1, \varepsilon'_2, \varepsilon'_3, \varepsilon'_4, \delta_1, \delta_2, \delta_3, \delta_4, V_1, V_2, V_3$, and V_4 = dielectric constant, loss angle, and volume of cell wall substance, air, bound water and free water, respectively

k = coefficient which varies from 0 to 1 and is determined experimentally or from theoretical premises.

If the value of $tg\delta$ of all components is below 0.1, Eq. (2.2) is simplified to:

$$tg\delta_m = \frac{V_1(\varepsilon'_1)^k tg\delta_1 + V_2(\varepsilon'_2)^k tg\delta_2 + V_3(\varepsilon'_3)^k tg\delta_3 + V_4(\varepsilon'_4)^k \cdot tg\delta_4}{V_1(\varepsilon'_1)^k + V_2(\varepsilon'_2)^k + V_3(\varepsilon'_3)^k + V_4(\varepsilon'_4)^k} . \tag{2.3}$$

If the dielectric parameters of wood with a moisture content at the fiber saturation point are known, it becomes possible to determine the dielectric parameters of moist wood with a moisture content exceeding the fiber saturation point using a simplified model. In this case, the wood should be regarded as a material composed of three components, namely wood with a moisture content equal to that at the fiber saturation point (W) (at 20 °C, W is approx. 30%), air, and free water.

For determination of the ε'_m of wood with a moisture content above the fiber saturation point, the corresponding formula can be derived based on the following assumptions. Free water interacts with the electric field independent of the wood, and therefore the increase in the free water in a unit volume correspondingly changes the value of the dielectric constant and of the loss tangent. Moisture is distributed throughout the whole volume of the wood, and the share of the dielectric constant propagated in the sample of free water is proportional to its relative volume V_4 and is equal to $V_4 \cdot \varepsilon'_w$.

When wood is moistened, the air is replaced by water. The volume of the dislodged air is equal to the volume of absorbed water, and the share of air in the dielectric constant of the sample decreases by the value of $V_4 \cdot \varepsilon'_b$. The dielectric constant of wood sample with a moisture content exceeding the fiber saturation point can then be found from the equation

$$\varepsilon'_m = \varepsilon'_p + V_4(\varepsilon'_w - \varepsilon'_b) = \varepsilon'_p + V_4(\varepsilon'_w - 1) , \tag{2.4}$$

where

ε'_w, ε'_p, and ε'_b = dielectric constant values of free water, wood with a moisture content equal to the fiber saturation point, and air ($\varepsilon'_b = 1$).

Table 2.1. Relative content of free water in wood V_4 ($t = 20\,°C$)

Moisture content (%)	Density (ρ_0, g/cm^3) 0.3	0.4	0.5	0.6	0.7	0.8
40	0.026	0.035	0.044	0.052	0.061	0.070
60	0.078	0.105	0.130	0.156	0.183	0.210
80	0.130	0.174	0.218	0.260	0.305	0.350
100	0.182	0.244	0.305	0.365	0.426	–
120	0.235	0.313	0.392	0.470	–	–

Equation (2.4) can be used if the relative content of free water in wood is known, related to the density and moisture content values of wood. The relative content of free water V_4 for the case in which the coefficient of the volumetric shrinkage is equal to 0.5% per 1% moisture content is given in Table 2.1.

The values of $tg\delta_m$ of wood with a moisture content above the fiber saturation point can be derived from:

$$tg\delta_m = tg\delta_p + V_4 \cdot tg\delta_w \,, \tag{2.5}$$

where

$tg\delta_p$ = loss tangent of wood at the moisture level equal to the fiber saturation point

$tg\delta_w$ = loss tangent of free water

V_4 = the relative volume of free water in the wood.

For the determination of the oven-dry wood dielectric constant, assuming that the wood is a two-component mixture consisting of cell wall substance and air, it is recommended employing for Eq. (2.1) a coefficient $k = 1/3$. In this case, Eq. (2.1) takes the form:

$$(\varepsilon'_m)^{1/3} = V_1(\varepsilon'_1)^{1/3} + V_2(\varepsilon'_2)^{1/3} \,. \tag{2.6}$$

The results of calculations using Eq. (2.6) show a good agreement with the experimental data. The error of calculation in using this formula of the dielectric constant of oven-dry wood with a density equal to 0.3–1.0 g/cm^3 in the frequency range from 10^2 to 10^{11} Hz is within + 5 to − 10% relative to the average empirical values.

2.2 Dielectric Parameters of the Cell Wall Substance and Its Elements

The cell wall substance is nonuniform and a compound dielectric consisting of cellulose, hemicelluloses, lignin, and extractives (resins, fats, tannins, etc.). Table 2.2 shows percentages of the main components relative to the total mass of the wood.

Table 2.2. Chemical composition of wood from some tree species as percentage of total mass (Borovikov and Ugolev 1989)

Species	Cellulose	Lignin	Hemicellulose		Ash	Extractive substances soluble in:	
			Pentosans (xylan)	Hexosans (mannan)		Ether	Water
Pine	51.9	20.5	11.2	9.3	0.2	2.6	0.6
Spruce	45.2	29.0	9.3	9.2	0.3	1.5	1.4
Larch (Siberian)	45.8	29.5	9.3	–	1.0	3.6	5.1
Oak	38.9	23.8	22.8	6.3	0.3	4.3	1.8
Birch	34.1	21.2	22.0	20.0	0.4	1.3	–
Poplar	39.8	18.8	23.5	8.3	0.3	6.8	2.1

Cellulose and hemicellulose are the polysaccharides which constitute about 70%–75% of the wood substance. These components determine the dielectric properties of the cell wall substance. Lignin is a mixture of irregularly ramified polymers based on phenyl propane units. Molecules of cellulose are composed of glucose residues with a degree of polymerization (Fig. 2.1). The polymerization degree of the wood cellulose is approximately 10000.

Cellulose, hemicelluloses, and lignin are polar polymers. Characteristic of them are electronic and ionic (atomic) types of polarization which are not connected with energy losses, as well as the dipole kind of polarization causing energy losses because of the considerable value of the relaxation time. These substances, which are considered to be high-molecular-weight agents with a number of polar groups, are subject to dipole-radical polarization under the action of an electric field. This kind of polarization is caused by the displacement of polar groups relative to the immovable parts of the macromolecules. Macromolecules of cellulose, hemicelluloses, and lignin under the influence of an electric field possess an ionic conductivity that increases due to the presence of other conductive substances in the cell walls. Later, we shall examine the

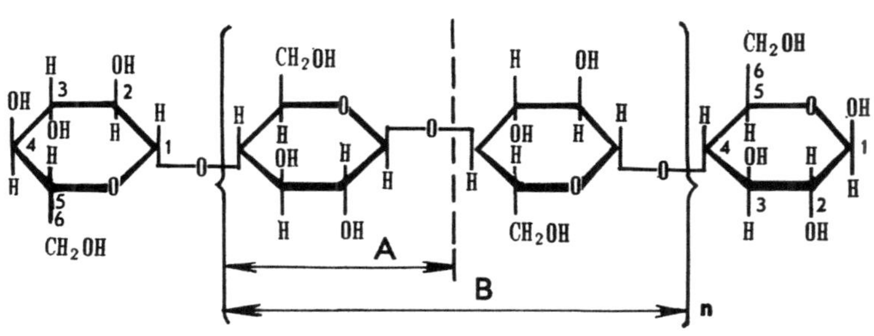

Fig. 2.1. Structural formula of cellulose: *A* β-D-glucose residue; *B* cellobiose

dielectric properties of the principal chemical components of the cell wall substance, i.e., cellulose, hemicelluloses, and lignin.

The process of cellulose polarization in an electric field is associated with a displacement and rotation of hydroxyl groups (–OH), methylol groups ($-CH_2OH$), as well as the molecules of cellulose. The character of these processes determines the dielectric properties of cellulose.

The dependence of $\varepsilon'_\perp$ and $tg\delta_\perp$ of oven-dry cellulose and of cell wall substance on the frequency in accordance with Pungs' data (1954) is given in Fig. 2.2. With the increase in frequency from 50 to 3×10^8 Hz at $t = 20\,°C$, the value of the dielectric constant decreases from 7.3–6.0 to 4.9–3.7, and the value of $tg\delta$ increases. This increase of $tg\delta$ occurs up to a frequency of 10^7 Hz and then decreases, the maximum value being equal to 0.07–0.08. This course of the curves of $\varepsilon'_\perp$ and $tg\delta_\perp$ of practically pure cellulose is very similar to those for the cell wall substance. The loss factor $\varepsilon''_\perp$ for both substances at 20 °C exhibits its highest value at a frequency of about 10^7 Hz. Cellulose shows higher values of $\varepsilon'_\perp$ and $tg\delta_\perp$ than the cell wall substance.

Each glucose residue in cellulose (Fig. 2.1) contains three hydroxyl groups. The presence of these groups is responsible for the polarizability of cellulose, as under the influence of an electromagnetic field the hydroxyl groups are capable of shifting relative to the entire molecular chain. This process results in the effect of a dipole-radical polarization. The dielectric constant (DC) of cellulose triacetate (which does not possess hydroxyl groups) at a frequency of 10^3 Hz and a temperature of 25 °C is equal to 4, while that of regenerated cellulose is equal to 7, which can be explained by the participation of the hydroxyl groups in the polarization of cellulose.

Cellulose contains β-D-glucosidic links. This is the reason why the properties of β-D-glucose determine the dielectric properties of cellulose. The results of measurement of ε' and $tg\delta$ of β-D-glucose are shown in Fig. 2.3 (Pungs 1954). There is clearly a difference of the curves in comparison with those of cellulose (Fig. 2.2). The maximum value of $tg\delta$ is shifted to the region of 10^8 Hz and manifests itself in the sharpness of the curve, while the curve of ε' at these frequencies is notably decreasing. This can be explained by the fact that glucose consists of simple identical molecules with the same relaxation time. Molecules

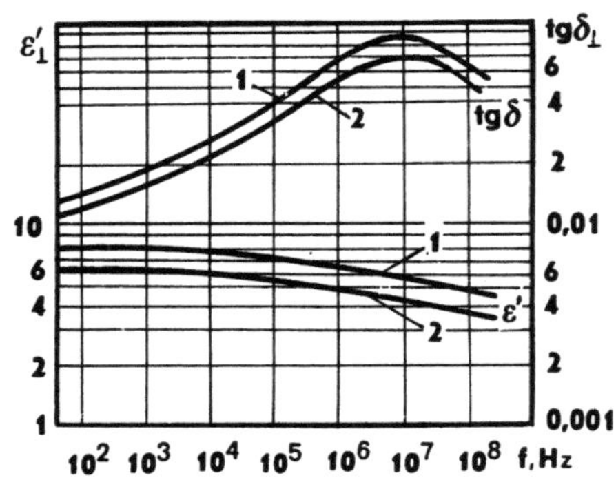

Fig. 2.2. Dependence of dielectric constant $\varepsilon'_\perp$ and loss tangent $tg\delta_\perp$ of oven-dry pure cellulose (*1*) and of the cell wall substance (*2*) on the frequency with the orientation of vector $\overline{E}$ perpendicular to the fiber direction (Pungs 1954); *f* frequency

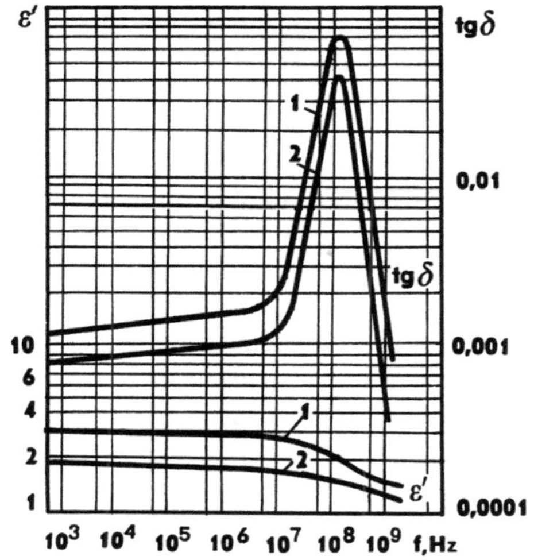

Fig. 2.3. Curves showing the dependence of ε' and $tg\,\delta$ of oven-dry glucose on the frequency at 20 °C (Pungs 1954): *1* pure glucose; *2* glucose 45% and air 55%

of cellulose are of various lengths and are composed of different numbers of glucose residues. This is the reason why the relaxation time is different, the actual time of relaxation being distributed around the most probable value of τ_m.

The cellulose fibers have crystalline and amorphous regions, the former amounting to 50%–60% in wood. The relaxation processes that take place in the moistureless cellulose under the influence of an electromagnetic field are also linked with the orientational polarization of the methylol dipoles (CH_2OH) in the amorphous regions. The dielectric properties of cellulose depend on the quantitative proportions of the crystalline ($\rho_c = 1.59\,g/cm^3$) and of noncrystralline ($\rho_a = 1.47\,g/cm^3$) regions. The relationship between the dielectric properties and the crystallinity of cellulose was studied by Norimoto (1976) who analyzed contributions by various investigators. Investigatioin of DC values of six kinds of cellulose showed that at 30 °C and at a frequency of 10^4 Hz, the value of $\varepsilon''_{||}$ of the crystalline region is equal to 8.60 and that of the amorphous region is 3.88. It was deduced theoretically the linear dependence between ε' and the volume fraction of crystallinity V_c in cellulose.

$$\varepsilon'_c = C_1 V_c + C_2 \,, \tag{2.7}$$

$$C_1 = 2\left(\rho_c - \rho_a\right)\frac{n-1}{\rho}\cdot\left(1 - \frac{n-1}{\rho}\rho_a\right), \tag{2.8}$$

$$C_2 = \left(1 + \frac{n-1}{\rho}\cdot\rho_a\right), \tag{2.9}$$

where

n = average refractive index
ρ_c, ρ_a = densities of crystalline and noncrystalline regions
ρ = the average density of the substance.

According to Norimoto et al. (1978), the values of the cellulose DC at 20 °C and at a frequency of 10^6 Hz are $\varepsilon'_{\perp} = 4.01$ and $\varepsilon'_{\parallel} = 4.25$ for crystalline and 7.83 and 8.06 for amorphous regions, respectively. At $t = -32$ °C and 10^3 Hz, the dry crystalline cellulose shows $\varepsilon'_{\parallel} = 3.1$ and the amorphous, 8.1.

With an increase of the duration of cellulose grinding in a ball mill, the DC of cellulose increases, which can be attributed to the diminution of the crystallinity. The decrease of the cellulose crystallinity index from 88 to 74% causes an increase in the DC of the cellulose with density equal to 1.2 g/cm^3 from 3.04 to 3.5 (at $t = -32$ °C and a frequency of 10^3 Hz).

The following conclusion can be drawn from the above, namely that in the frequency range from 10^3 to 10^6 Hz at a temperature ranging from + 30 to − 32 °C, $\varepsilon'_{\parallel}$ of the noncrystalline fraction of the cellulose substance lies within 8.1–8.6, and $\varepsilon'_{\parallel}$ of the crystalline fraction is equal to 3.1–4.25. The relation $\varepsilon'_{\parallel}/\varepsilon'_{\perp}$ of cellulose is equal to 1.03–1.06.

The coefficients of refraction n of natural crystalline cellulose in longitudinal and in transverse directions at optical frequencies are 1.618 and 1.543, respectively (Hermans 1949). According to the well-known Maxwell formula $n^2 = \varepsilon'$ the DCs of crystalline cellulose at a frequency of 10^{15} Hz are $\varepsilon'_{\parallel} = 2.62$ and $\varepsilon'_{\perp} = 2.38$.

Direct measurements of the dielectric parameters of cellulose are extremely difficult. They can be determined by measurement of the dielectric parameters of mixtures containing cellulose, air, and nonpolar substances whose dielectric parameters are known. The parameters of cellulose can then be calculated on the basis of the compound dielectric theory, e.g., using formulas similar to Eqs. (2.1) and (2.3). This may be an explanation why the data obtained by some investigators at various frequencies differ from each other by 12–34%.

Proceeding from empirical data pertaining to the relations between the dielectric parameters of the cell wall and those of cellulose ε'_c, I have calculated values of ε'_c and $tg\delta_c$ at frequencies of 10^2–10^{10} Hz. The results of these calculations are given in Table 2.3.

The dependence of the DC and the loss factor of cellulose with a density of 1.3 g/cm^3 upon the temperature is shown in Fig. 2.4 (Norimoto 1976). With an increase of temperature, the value of ε' increases, while ε'' passes its maximum value which, on reduction of the frequency, shifts in the direction of lower temperatures. The relaxation process occurs in dry cellulose also at negative

Table 2.3. Dielectric parameters of wood cellulose at 20 °C

Frequency (Hz)	10^2	10^3	10^4	10^5	10^6	10^7	10^8	10^9	10^{10}
$\varepsilon'_{c\perp}$	7.1	7.0	6.9	6.4	6.0	5.5	4.9	3.8	3.6
$tg\delta_{c\perp}$	0.018	0.022	0.031	0.045	0.064	0.080	0.070	0.050	0.047
$\varepsilon''_{c\perp}$	0.128	0.154	0.214	0.288	0.384	0.440	0.343	0.190	0.169
$\varepsilon'_{c\parallel}$	7.5	7.4	7.3	6.8	6.4	5.8	5.2	4.1	3.9

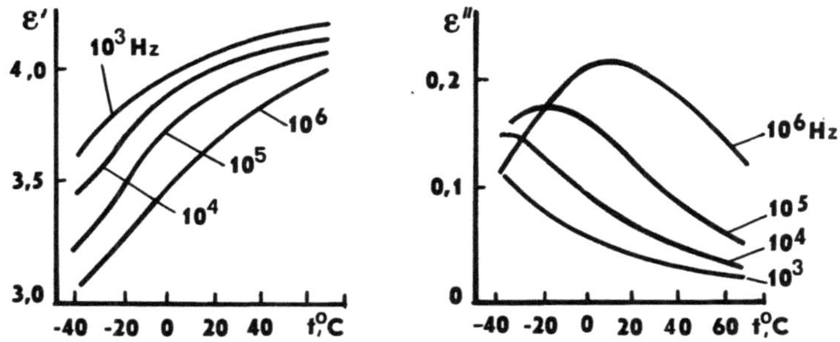

Fig. 2.4. Temperature dependence of ε′ and ε″ of cellulose with a density of 1.3 g/cm³ and a moisture content of 0.2% (Norimoto 1976)

temperatures. This process is connected with orientational polarization of the methylol groups in the amorphous regions.

At a frequency of 50 Hz, the maximum of the loss factor of cellulose occurs at a temperature of − 80 °C. The peak value of the loss factor takes place at a frequency of $\omega_0 = 1/\tau_0$, which is called the relaxation frequency. Proceeding from the fact that the peak of ε_c'' of cellulose at 20 °C is at a frequency of 10^7 Hz and that at − 80 °C it occurs at 50 Hz, the relaxation time for cellulose at these temperatures can be found from:

$$(\tau_0)_{20} = 1/\omega_0 = 1/2\pi f = 1.59 \times 10^{-8}\,\mathrm{s}$$

$$(\tau_0)_{-80} = 3.18 \times 10^{-3}\,\mathrm{s}\,.$$

Thus, the DC of cellulose, which is the main component of the cell wall, at a temperature of 20 °C and within the frequency range from 10^2 to 10^{15} Hz changes from 7–7.5 to 2.4–2.6. The loss tangent will peak at a frequency of 10^7 Hz. The dielectric properties of the cell wall substance are to a great extent determined by the properties of cellulose.

The dielectric properties of one of the main kinds of hemicelluloses, viz. glucomannan, are close to those of cellulose. With the increase of frequency from 3×10^2 to 10^6 Hz and at $t = 20$ °C, the loss factor of glucomannan increases from 0.06 to 0.15, and at $t = -20$ °C it is within 0.04–0.13 (Norimoto 1976). The maximum values of the loss factor of glucomannan as a function of temperature and at various frequencies are near to those of cellulose. This coincidence can be explained by the presence of methylol groups in both glucomannan and cellulose. Xylan does not contain methylol groups, and therefore its dielectric properties are of lower values. The loss factor of xylan at 20 °C in the frequency range 3×10^2–10^6 Hz changes within 0.05–0.07, and at $t = -20$ °C it is equal to 0.03–0.04.

Lignin shows lower dielectric properties in comparison with those of cellulose despite the presence of polar groups such as $-OH$ and $-CH_2OH$. Table 2.4 shows the dielectric values for lignin at 20 °C at various frequencies.

Table 2.4. Dielectric parameters of dry lignin at 20 °C

	ρ (g/cm^3)	Frequency (Hz)							
		50	10^2	10^3	10^4	10^5	10^6	3×10^6	
ε_l'	1.34	4.25	4.24	4.19	4.13	4.06	3.96	3.89	Norimoto 1976
ε_l'	1.23	–	3.97	3.93	3.87	3.81	3.72	–	Personal
$tg\delta_l$	1.23	–	0.004	0.006	0.007	0.009	0.013	–	observations

The DC of lignin of any density and at 20 °C within the frequency range from 50 up to 3×10^6 Hz can be calculated by using the following equation (Venkateswaran 1972):

$$\varepsilon_l' = (\rho/1.34)(\varepsilon_l' - 1) + 1\,, \tag{2.10}$$

where

ε_l' = DC of oven-dry lignin with a density of 1.34 g/cm^3 according to Table 2.4.

The loss factor of lignin has a complicated dependence on frequency and moisture content. Lignin with a density equal to 1.23 g/cm^3 and a moisture content of 0.2% shows a maximum value of the loss factor ε'' equal to 0.057 at a frequency of 10^6 Hz and temperature of -20 °C, while at a frequency of 10^5 Hz the maximum value of ε'' is at $t = -50$ °C. Figure 2.5 shows the loss factor of lignin as a function of temperature and frequency.

Some inorganic constituents are also present in the cell wall. However, because of their small quantities (averaging about 0.3%), these substances have little influence on the dielectric parameters of wood.

The substances forming wood, namely cellulose, hemicelluloses, and lignin, are polar dielectrics that contribute to the dielectric parameters of the cell wall substance. Their influence depends upon their own properties and quantities. It is an extremely complicated task to calculate theoretically the dielectric parameters of cell wall substances. Norimoto (1976; Norimoto et al. 1978) made

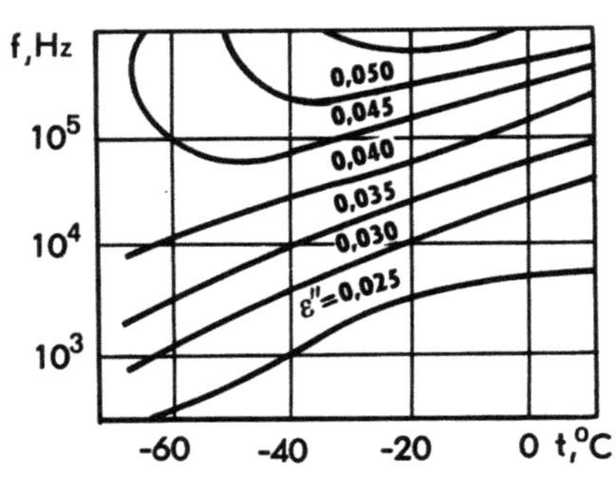

Fig. 2.5. Loss factor ε'' of lignin with a density of 1.23 g/cm^3 and a moisture content of 0.2% as a function of temperature and frequency (Norimoto 1976)

a valuable contribution to solving this problem. He theoretically calculated the value of the wood DC in relation to the cell structure, constructed mathematical models that linked the dielectric parameters of the chemical components of wood with the DC of the cell walls and computed the DCs of earlywood and latewood and of the rays. According to Norimoto (1976), the DC of tracheid walls ε'_{cw} can be derived from the equation:

$$1/\varepsilon'_{cw} = \cos^2\theta/\varepsilon'_{cw1} + \sin^2\theta/\varepsilon'_{cw2}\ , \qquad (2.11)$$

where

ε'_{cw1}, ε'_{cw2} = the DC of the cell wall in parallel and perpendicular directions to the microfibrils, respectively, as given in Table 2.5

θ = the microfibrillar angle relative to the longitudinal axis of a cell.

Table 2.6 shows some calculated DCs for the cell walls of tracheids and of rays in longitudinal, radial, and tangential directions at a frequency of 10^6 Hz and a temperature of 20 °C.

The DC and loss tangent of a cell wall substance can be determined experimentally by measurements of parameters of a composition that includes the cell wall substance and nonpolar substances with known dielectric properties and by subsequent calculation of ε' and $tg\delta$ from the formulas of the compound dielectric theory.

Analysis of experimental data concerning dielectric parameters of the cell wall by using Cole-Cole diagrams has shown that its limiting low-frequency DC at 20 °C is equal to $\varepsilon'_{\perp s} = 6.8$, whereas the high-frequency one is $\varepsilon'_{\perp\infty} = 2.9$.

Table 2.5. Dielectric constants of dry cell walls in directions parallel (ε'_{cw1}) and perpendicular (ε'_{cw2}) to the microfibrills at 20 °C (according to data from Norimoto 1976)

Frequency (Hz)	3×10^2	10^3	10^4	10^5	10^6	3×10^6
ε'_{cw1}	7.3	7.1	6.9	6.4	5.3	4.5
ε'_{cw2}	5.7	5.6	5.5	5.2	4.8	4.3

Table 2.6. Dielectric constants of cell walls of coniferous wood elements at a frequency of 10^6 Hz and at 20 °C (Norimoto 1976)

Direction of vector $\bar{E}$	Earlywood		Latewood		Rays
	Radial wall	Tangential wall	Radial wall	Tangential wall	
L	5.09	5.14	5.23	5.22	4.64
R	4.84	4.73	4.77	4.76	4.64
T	4.73	4.80	4.76	4.78	4.64

Within these limits, the DC of the cell wall monotonously diminishes with increasing frequency (see Fig. 4.4, $\rho_0 = 1.53\,g/cm^3$). The study of the DC of cell walls of earlywood and latewood and of ray cells has proved (Table 2.6) that the ε' of the cell wall substance of latewood and earlywood do not differ significantly from each other, but the DC of ray cell walls is lower than that of wood by 4–12%. The DC of the cell wall substance in the longitudinal direction is higher than that in the transverse direction by 5–25%.

The loss tangent of the cell wall in the transverse direction at 20 °C and at a frequency of 10^2 Hz is equal to 0.014. With increasing frequency, it assumes a maximum value equal to 0.06 at 10^7 Hz and then decreases to 0.039 at 10^{11} Hz (Fig. 4.5, $\rho_0 = 1.53\,g/cm^3$). In the longitudinal direction, $tg\delta$ is higher than in the transverse direction by 1.25–1.9 times (depending on the frequency). The loss factor changes within a range of 0.09–0.34, the peak value being at 10^7 Hz (Fig. 4.6, $\rho_0 = 1.53\,g/cm^3$).

Specific values of the dielectric parameters of the cell wall in the frequency range from 10^2 to 10^{11} Hz at 20 °C and with the vector of electric field strength oriented in the transverse direction are given in Appendix 1.

The DC of the cell wall increases with increasing temperature (Fig. 2.6a); the dependence is nearly linear. The dependence of the loss factor on the temperature at different frequencies is shown in Fig. 2.6b. The maximum value of $\varepsilon''_{cw\perp}$ at a frequency of 10^6 Hz is reached at $t = 20$ °C, while at frequencies of 10^2–10^4 Hz, the peak shifts to the negative regions of temperatures below − 40 °C. At frequencies equal to 10^8–10^{10} Hz, the maximum value of $\varepsilon''_{cw\perp}$ is at temperatures above 100 °C.

The relative contents of cellulose, hemicelluloses, and lignin can be quite different in different wood species. Wood structure can also vary from one tree species to another. However, the dielectric properties of the cell wall substance

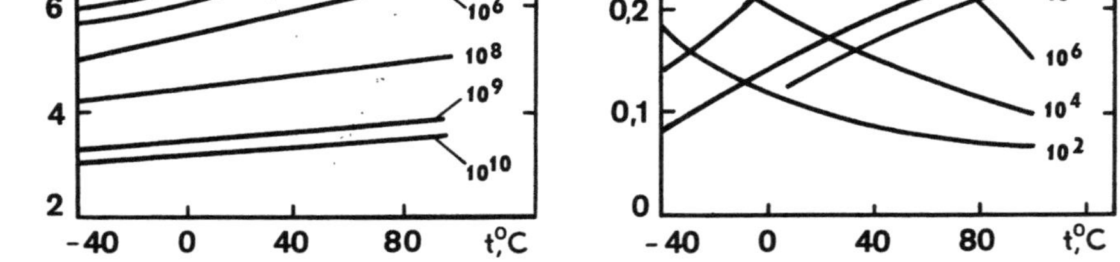

Fig. 2.6a, b. Curves showing dependence of dielectric constant $\varepsilon'_{cw\perp}$ (**a**) and loss factor $\varepsilon''_{cw\perp}$ (**b**) of dry cell wall substance on temperature

Table 2.7. Dielectric parameters of cell wall substance at temperatures ranging from − 40 to 100 °C

Frequency (Hz)	Temperature											
	− 40 °C		− 20 °C		20 °C		40 °C		60 °C		100 °C	
	$\varepsilon'_{cw\perp}$	$tg\delta_{cw\perp}$	$\varepsilon'_{cw\perp}$	$tg\delta_{cw\perp}$	$\varepsilon'_{cw\perp}$	$tg\delta_{cw\perp}$	$\varepsilon'_{cw\perp}$	$tg\delta_{cw\perp}$	$\varepsilon'_{cw\perp}$	$tg\delta_{cw\perp}$	$\varepsilon'_{cw\perp}$	$tg\delta_{cw\perp}$
10^2	5.9	0.031	6.1	0.025	6.5	0.014	7.2	0.013	7.5	0.011	7.8	0.008
10^3	5.8	0.040	6.0	0.032	6.4	0.018	7.0	0.016	7.4	0.014	7.7	0.009
10^4	5.7	0.048	5.9	0.040	6.3	0.025	6.9	0.022	7.2	0.019	7.6	0.013
10^5	5.4	0.059	5.5	0.052	5.9	0.037	6.5	0.031	6.8	0.025	7.1	0.015
10^6	5.0	0.056	5.2	0.055	5.5	0.053	6.1	0.046	6.3	0.038	6.6	0.023
10^7	4.7	0.047	4.8	0.054	5.0	0.067	5.3	0.064	5.5	0.062	5.6	0.056
10^8	4.2	0.035	4.3	0.038	4.5	0.061	4.8	0.066	4.9	0.071	5.0	0.081
10^9	3.3	0.028	3.4	0.030	3.5	0.049	3.6	0.053	3.7	0.057	3.8	0.065
10^{10}	3.1	–	3.2	–	3.3	0.043	3.4	0.048	3.5	0.053	3.6	0.064

are considered to be independent of the tree species because the errors inherent in their experimental determination exceed by far the influence of any differences in chemical composition and in the wood structure. Table 2.7 gives the values of the DC and the loss tangent of the cell wall substance calculated in terms of empirical data for a frequency range from 10^2 to 10^{10} Hz and at temperatures from -40 to $+100$ °C.

2.3 Dielectric Characteristics of Air, Free and Bound Water, and Ice

The dielectric constant (DC) of dry air in the temperature range from -50 to $+80$ °C changes from 1.0007 to 1.0004. With an increase of relative humidity up to 90–100%, the DC of air rises to 1.002–1.003. The changes in DC of air under ordinary conditions is therefore somewhere around the third digit after the decimal point. This is the reason for assuming in practical calculations that the DC of humid air is equal to unity and independent of the frequency. In the absence of collision ionization conditions, air is a perfect dielectric, and its $tg\delta$ is considered equal to zero. Moisture values of wood can change from zero to 250%. It is customary to distinguish between several levels of moisture content. These levels represent qualitative differences among the bonds between moisture and wood (Chudinov 1984; Ugolev 1986; Skaar 1988) and also determine the dielectric properties of the moist wood. Wood dried at 103 ± 2 °C is called oven-dry wood ($W = 0\%$). The maximum value of the cell wall moisture content W_f, that can be achieved if wood is moistened directly in water is called the fiber saturation point. This state is characterized by an equilibrium between the cell wall moisture content and the water located in the cell cavities. For wood species growing in temperate zones, a mean value of $W_f = 30\%$ is usually accepted. However, in some species (especially in those with a low wood densityi) deviations from this value can be considerable.

The maximum value of the cell wall moisture content, W_g, is called the hygroscopicity limit. This state is achieved when vapor sorption takes place. It is characterized by the absence of water in the cell cavities and by an equilibrium between the cell wall moisture content and that of air, the humidity of which is approaching the saturation point. Whereas a change of temperature does not affect the fiber saturation point value, W_f, the hygroscopicity limit, W_g, decreases with increasing temperature, and at 100 °C it is equal to 19–20%. Wood contains two kinds of water, namely free and bound water. Free water, which is inside the cell cavities, is kept there mechanically and by the forces of mutual capillary interaction. The free water is assumed to be inside the macrocapillaries, i.e., in pores and capillary tubes with a radius exceeding 10^{-5} cm (Chudinov 1984).

Water present in capillary tubes of a lesser diameter is related to the bound water. The bound (or hygroscopic) water resides in the walls of cells. This water is distinguished into two groups: adsorption water and capillary-condensed water. The water sorbed by surfaces of the microfibrils is called adsorption water. This water forms intermicrofibrillar interlayers and is bound to the cell wall by physical and chemical forces. Adsorption water is separated into monomolecular and polymolecular water. Capillary-condensed water occurs in the cell wall pores and is bound to the wood by capillary forces.

Thus, three forms of bound water are distinguished in wood, namely monomolecular, polymolecular, and capillary-condensed in the order of decreasing bond energies. Definite borders between these fractions of bound water are not established, and most likely they are "eroded," as the transfer from one fraction to another envelops several layers. Changes in the nature of the water in the process of its transition into the bound state and within each fraction involve changes of the properties of water and depend upon the energy of its interaction with the surface of the material.

Monomolecular layers of water interact most intensely with the cell wall. As the number of layers increases, the interaction between the adsorbed moisture and the wood diminishes, which manifests itself in a decrease in the heat of wetting. According to Chudinov and Andreev (1978), the maximum quantity of adsorption moisture over a temperature range of 0–20 °C is about 23%, at 100 °C it is near 18%, and at − 50 °C it is equal to 13%. The properties of the capillary-condensed moisture are between those of adsorption and free water. Chudinov (1984) has shown that the moisture-holding capacity of cell walls in relation to adsorption water depends to only a small degree on the tree species, whereas the cell wall moisture capacity in relation to the capillary-condensed water depends completely upon the anatomy of the wood and the ultrastructure of the cell wall.

For an estimation of the influence of different forms of water on the dielectric properties of wood, the following four ranges of moisture content can be conditionally established:

1. Monomolecular moisture – from oven-dry wood to 5% moisture content;
2. Polymolecular moisture – from 5% up to 18–23%;
3. Capillary-condensed moisture – from 18–23% up to the fiber saturation point content (30%);
4. Free water – moisture content above the fiber saturation point ($W > 30\%$).

At temperatures below zero, a part of the moisture in wood occurs in the form of ice, while another part is in a nonfreezing condition. The content of the nonfreezing moisture in the frozen wood depends on the moisture content and on the negative temperature. The freezing process is influenced by such factors as the porosity of the wood, the difference in the properties of the water in macroscopic and microscopic interstices, interaction forces between water and the capillary walls, minerals and organic substances dissolved in the water, and other factors.

The process of wood defrosting is conventionally divided into three temperature ranges of the phase transition (Chudinov 1968):

1. A range of considerable phase transformations occurring at a temperature of approximately 0 °C; this phase entails a defrosting of free water in the wood.
2. A range with insignificant phase transformations occurs at negative temperatures above − 50 °C. In this temperature range, the thawing out of the hygroscopic moisture of the wood occurs.
3. The range of frozen wood in the temperature interval below − 50 °C.

The change of the aggregate state of moisture in the wood does not cease entirely at temperatures below − 50 °C. The phase changes continue also at lower temperatures, but the amount of water that participates in the phase transformations is so inconspicuous that for practical purposes it is not taken into consideration.

Investigations using nuclear magnetic resonance spectroscopy have shown that all water in wood loses its liquid properties at a temperature of -80 ± 3 °C (Chudinov 1968). However, this does not mean that all moisture in the wood is converted into ice. According to experimental data, 5–6% of the moisture in wood occurs in monomolecular layers. The ratio of the heat of hydration to the quantity of the adsorbed moisture, which is equal to 335 kJ/kg, coincides with the latent melting heat of the ice. This means that in the process of binding with wood the adsorbed moisture has liberated a quantity of energy that is exactly equal to the energy which would be liberated at the transformation of this water into ice. This is the reason why such common definitions as "freezing" and "thawing out" cannot be applied to this kind of water.

Freezing of hygroscopic moisture present in the microcapillary tubes is characterized by a spectrum of temperature values. Chudinov (1968, 1984) has estimated the defrosting temperatures for this moisture in wood with a moisture content exceeding the fiber saturation point. The spectrum was computed on the basis of experimental and theoretical calculations that can be expressed by the following formula:

$$w = 12 + 19.5\exp(0.055t)\,, \tag{2.12}$$

where

w = relative quantity of nonfreezing moisture in the wood (%)
t = temperature (°C).

This formula is valid only for heating of wood in the temperature interval from − 50 °C up to 0 °C. Because of hysteresis effects in the processes of frosting and defrosting of the wood, the relation $w = \phi(t)$ for the freezing process is somewhat different.

The ratio between the quantity of ice and of nonfreezing moisture in the wood is the factor that determines the change in dielectric properties of moist wood at negative temperatures.

When the moisture content exceeds the fiber saturation point, the free water molecules interact with the electromagnetic field independently of the wood and its bound moisture. The volume of free moisture in the wood may be large enough to change considerably the dielectric parameters of the wood. The dielectric properties of free moisture are well-known, although there are some anomalies. Water is constantly dissociating and is subject to the spontaneous formation of such ions as hydroxyl (OH–) and hydronium (H_3O^+); due to them, water possesses electric conductivity features. At the same time the the molecules of water tend to bind mutually and each molecule is linked with four adjacent ones. As a result, a coherent spatial system, similar to a gigantic polymer, is formed. Separate structures of various kinds can be generated in this system. Water can preserve such structures for some time, which determines many of its anomalous properties.

The static DC of water at 20 °C is about 80, and at 200 °C it is equal to 34.6. With a rise in frequency from 0 to 10^9–5×10^9 Hz and at a constant temperature, changes in the DC of water are insignificant (Fig. 2.7a). In the same frequency

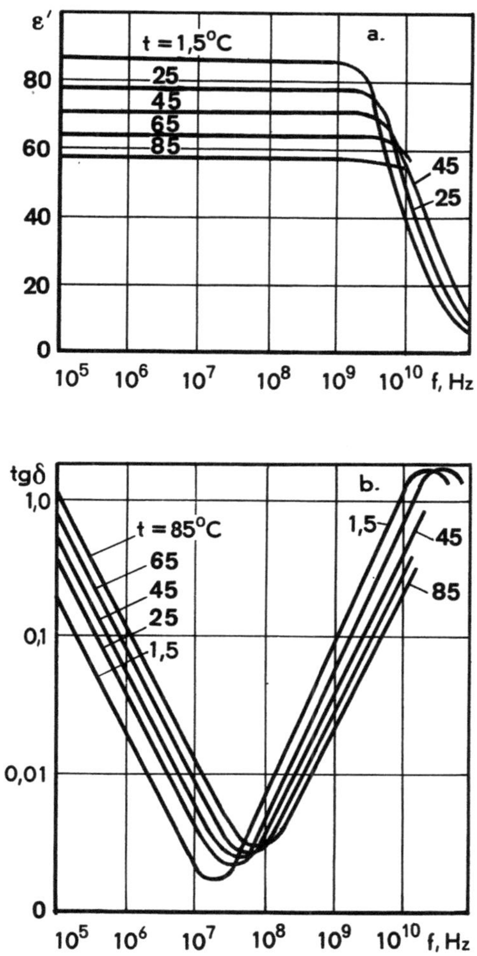

Fig. 2.7a, b. Dependence of the dielectric properties of free water on frequency (von Hippel 1954): **a** dielectric constant; **b** loss tangent

range, a rise in temperature causes a decrease in the DC. In the frequency range from 10^9 to 10^{10} Hz, the DC of water abruptly drops, and at a frequency of 3×10^{11} Hz it is equal to 5.7 (at $t = 20\,°C$). At a frequency of about 10^{10} Hz, the rise in temperature results in an increase in the DC. The DC of water is near its static value until the frequency is sufficiently low so that the dipoles can follow the changes in the field. When the frequency rises so high that the molecules cannot follow the field changes, the DC decreases and its value approaches that determined by electronic polarization.

The value of the loss tangent (Fig. 2.7b) diminishes with an increase of frequency above 10^5 Hz, the minimum value being between 10^7 and 10^8 Hz, and then increases again with a maximum at a frequency of about 6×10^{10} Hz, at $t = 20\,°C$ (Rosenberg 1979), after which it decreases once more. At frequencies upto 10^7–10^8 Hz, the loss tangent has a positive temperature coefficient, while after passing the minimum value, it changes into a negative one.

The frequency dependence of the dielectric properties of water can be explained by a mutual superimposition of the relaxation effects and the conductivity of the direct current. At frequencies below 10^7 Hz, the losses in water are caused mainly by its conductivity. When the frequency rises above 10^8 Hz, the relaxation losses are higher than those caused by the conductivity of the direct current. The maximum value of the coefficient of relaxation losses of water at 20 °C lies at about 1.5×10^{10} Hz (Fig. 2.8), and at 40 °C it is at a frequency equal to 2.4×10^{10} Hz (Rosenberg 1979). The relaxation time of the molecules of water at 20 °C is 1.06×10^{-10} s, and at 40 °C it is 0.66×10^{-10} s.

The dielectric parameters of water at positive temperatures and those of supercooled water at negative temperatures that are required for further calculations are given in Table 2.8. This table is compiled from data by von Hippel (1954) and Rosenberg (1979). The dielectric properties of water change considerably with changes in frequency and temperature. In most cases, they exceed those of the cell wall substance. This is the reason why changes in ε' and $tg\delta$ of

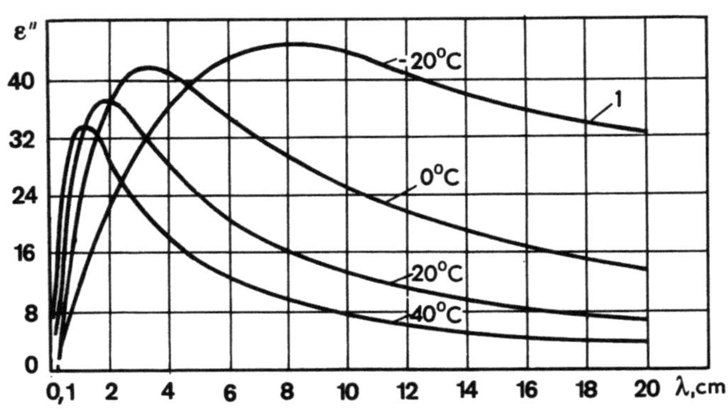

Fig. 2.8. Free water loss factor as a function of wavelength λ in a frequency range from 1.5×10^9 to 3×10^{11} Hz (Rosenberg 1979); 1 supercooled water

Table 2.8. Dielectric parameters of water (von Hippel 1954 and Rosenberg 1979)

Frequency (Hz)	Parameters	Temperature							
		− 30 °C	− 20 °C	− 10 °C	0 °C	20–25 °C	40–45 °C	65 °C	85 °C
10^5	ε'	–	–	–	87	78.2	71.5	65	58
	$tg\delta$	–	–	–	0.19	0.4	0.55	0.9	1.24
10^6	ε'	–	–	–	87	78.2	71.5	64.8	58
	$tg\delta$	–	–	–	0.019	0.022	0.059	0.086	0.124
10^7	ε'	–	–	–	87	78.2	71	64	58
	$tg\delta$	–	–	–	0.002	0.0046	0.005	0.009	0.0125
10^8	ε'	–	–	–	87	78	71	64	58
	$tg\delta$	–	–	–	0.007	0.005	0.004	0.0035	0.003
10^9	ε'	–	–	–	86	77	71	64	57
	$tg\delta$	–	–	–	0.102	0.06	0.034	0.025	0.019
2.4×10^9	ε'	49	69	79	82	77	71	64	57
	$tg\delta$	0.92	0.62	0.40	0.25	0.13	0.09	0.06	0.04
5.8×10^9	ε'	21	37	54	65	68	66	62	55
	$tg\delta$	1.62	1.2	0.80	0.59	0.31	0.22	0.17	0.13
10^{10}	ε'	10.6	17.6	29	38	55	59	59	54
	$tg\delta$	2.03	1.75	1.33	1.03	0.54	0.40	0.32	0.26
3×10^{10}	ε'	6.1	7.0	8.9	12.3	23.2	35.6	–	–
	$tg\delta$	1.2	1.66	1.89	1.84	1.37	0.94	–	–
10^{11}	ε'	5.55	5.64	5.82	6.16	7.5	10.1	–	–
	$tg\delta$	0.41	0.63	0.90	1.19	1.61	1.7	–	–

Table 2.9. Dielectric parameters of ice at temperatures from − 11 °C to − 12 °C (Auty and Cole 1952; von Hippel 1954)

Parameters	Frequency (Hz)												
	0	50	10^2	10^3	2×10^3	5×10^3	10^4	10^5	10^6	10^7	3×10^9	10^{10}	∞
ε'	95	95	94.6	83	60.6	22.5	9.5	4.8	4.15	3.7	3.2	3.17	3.08
$tg\delta$	–	0.021	0.038	0.365	0.728	1.68	2.4	0.8	0.12	0.018	0.0009	0.0007	–
ε''	–	2.0	3.6	30.3	44.1	37.8	22.8	3.84	0.498	0.067	0.0029	0.0022	–

wood with moisture contents exceeding the fiber saturation point are determined mainly by the dielectric parameters of the free water and by its volume in the wood. The dielectric properties of bound water are still poorly understood. Most investigators draw their conclusions concerning the dielectric properties of bound water by comparing results obtained by measuring the dielectric properties of over-dry material and those of materials with differing moisture contents. In these studies, the influence of the cell wall is taken into consideration to some extent. The major error is introduced into the results and the conclusions by the fact that the mass of bound water in the sample is several times less than the mass of the sample itself. The estimations of the dielectric parameters of the bound water are therefore largely qualitative. According to the data reported by various authors (Zhilenkov 1957, 1970; Eme 1967; Nekrasov 1979; Benkova 1987; Benkova and Chudinov 1990), values for the DC of bound water differ extensively, namely, from 2 to 77. While free water and ice are characterized by a single relaxation time at a certain temperature, the adsorption moisture is notable for its wide spectrum of relaxation times.

Monomolecular moisture possesses the most stable bonds with the cell wall substance. Its molecules exhibit limited mobility under the action of an alternating electric field. This is the reason why, with an increase of moisture content from 0 to 4–6%, the rise of DC and $tg\delta$ of wood is insignificant. The layers of polymolecular moisture possess greater mobility and higher dielectric parameters, and as a result they contribute considerably to the dielectric parameters of moist wood.

According to Benkova and Chudinov (1990), who investigated the dielectric relaxation the water in wood at temperatures from + 20 to − 150 °C, the reorientational mobility of molecules of "multilayer" water is two orders lower and of "monolayer" water is four orders lower than that of free water molecules. The reorientational mobility of "monolayer" water molecules is one to two orders higher than that of the molecules of crystalline ice, and the mobility of "multilayer" water molecules is three to five orders higher than that of ice molecules. The conclusion can be deduced that the dielectric parameters of the adsorption moisture of wood lie in a region between the parameters of free water and those of crystalline ice, at the corresponding frequency and temperature.

Mutual interactions of bound water and wood as well as the complexity of the interactions between the composition of the wood and water on the one side and the electromagnetic field on the other side do not permit us to obtain either experimentally or theoretically separate data characterizing the parameters of the bound water. Therefore, for analyses of the dielectric properties of wood with a moisture content below the fiber saturation point, the wood should be considered as a uniform system without separation of the cell wall substance from the bound water.

At negative temperatures, the free water and a part of the bound water turn into ice, which cannot but affect the dielectric parameters of the wood. The dielectric properties of ice differ significantly from those of water. The static DC of ice rises with a lowering of the temperature, and at − 0.1 °C it is equal to 91.5, at − 11 °C to 95, at − 32 °C to 100, and at − 66 °C to 133 (Auty and Cole 1952). The limiting high-frequency DC does not depend on temperature (down to − 66 °C) and remains at 3.1. The character of the dependence of ε' and $tg\delta$ on frequency and temperature according to Netushil et al. (1959) is shown in Fig. 2.9. The dielectric parameters of ice at various frequencies are given in Table 2.9.

The change of the DC of ice at frequencies exceeding 10^9 Hz is comparatively small, and at temperatures from − 10 to − 20 °C the DC of ice is equal to 3.1–3.2. The loss tangent has peak values at temperatures from − 10 to − 50 °C and is equal to 2.2–2.6 in the frequency range 10^3–10^4 Hz. At higher frequencies, the value of the loss tangent diminishes and drops to 0.0007–0.0009 in the super high frequency range.

The activation energy for the process of ice polarization is 37.7 kJ/mol, whereas for water it is 16.8 kJ/mol.

The influence of ice on the DC of wood drops abruptly with increasing frequency up to 10^5 Hz and then decreases monotonously. The influence of ice

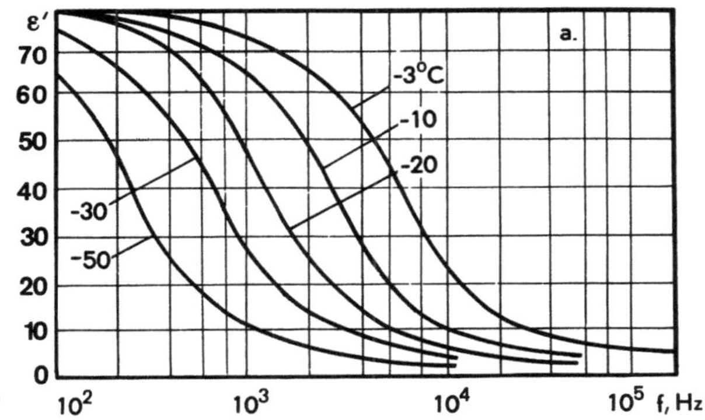

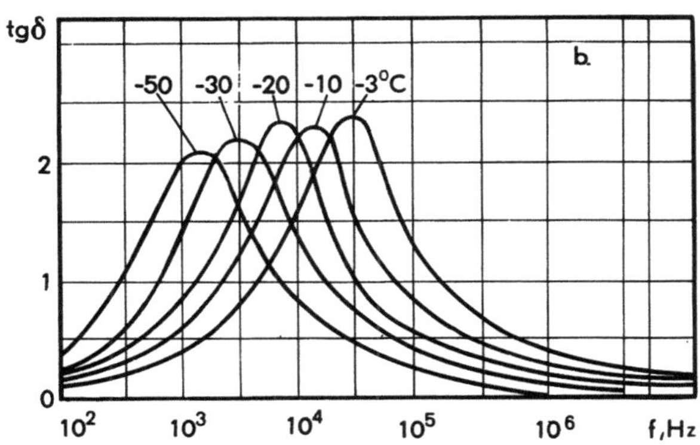

Fig. 2.9a, b. Dependence of ε' and $tg\delta$ of ice on the frequency (Netushil et al. 1959)

on the wood loss tangent is notable in the frequency range up to 10^7 Hz. At super high frequencies, the influence of ice on $tg\delta$ is inconspicuous and can be neglected.

It is worthwhile noting that wood at negative temperatures may still contain some unfrozen, supercooled moisture. The DC of this water at super high frequencies is from 2 to 25 times higher than that of ice, and its loss factor is several orders higher than the loss factor of ice (Table 2.8). This circumstance greatly affects the properties of the wood and causes a considerable spread in the data obtained.

The above survey shows that substances that occur in dry and moist wood may possess a great variety of dielectric parameters that depend on frequency and temperature. This variety of properties is determined by different kinds of electric processes caused by interactions with external electromagnetic fields on the molecular level. The scope of influence of the dielectric parameters of these substances on the dielectric properties of wood is determined by their own properties, their relative quantities, and the extent of their mutual interactions. The knowledge of these factors makes it possible to estimate the influence of every substance on the dielectric properties of wood, to predict the behavior of wood under the specific conditions of an electromagnetic field action, and to choose the method of measuring the dielectric characteristics of wood.

Wood is a very complicated material, and in many cases it is impossible to calculate theoretically its dielectric parameters with the accuracy required by practice. This explains why reliable data on the dielectric properties of wood can be obtained only by proper experimentation.

Chapter 3

Measurement of the Dielectric Properties of Wood

Measurements of the dielectric characteristics of wood includes the following:

1. The choice of the measurement technique that makes it possible to obtain the required parameters with the necessary accuracy.
2. Preparation of the samples for the tests.
3. Measurement of electric or nonelectric resultants which directly or indirectly characterize the wood properties with subsequent use of the data for calculation of the required parameters.
4. Definition of the rules governing the changes of the wood properties under the influence of the following factors:
   a) Changes of frequency, electric field strength, and field orientation relative to the principal axes of wood anisotropy
   b) Temperature and humidity values
   c) Physical and chemical actions exerted on the wood samples (pressure, radiation, chemical treatment, etc.).
5. Statistical data processing and presentation of the results of the measurements in forms convenient for use.

Methods of definition of the dielectric parameters and measurement techniques at various frequencies are described in the works by von Hippel (1954), Netushil et al. (1959), Brandt (1963), Bergmann (1969), Kasarnovskiy and Tareev (1980), and in some other special studies. The standard methods of measurement of the dielectric constant (DC) and loss tangent have been developed mainly for checking electric insulating materials, but wood is quite different. Wood is a multicomponent, anisotropic material which can keep moisture in the range from zero to 250%. Besides, wood changes its dimensions when it is moistened or dried. It also has other characteristics depending on the tree species and on the density of the wood, and its variability is large. All the above complicate the measurement techniques and introduce into the process a number of errors. Standards for measurements of the electric characteristics of wood and wood-based materials have not been worked out yet, and usually the general methods of corresponding test procedures for solid dielectrics are used for these purposes. Important complications arise in the process of the preparation of wood samples and preserving their specified moisture level during the test procedure, as a change of moisture content entails changes in the sample dimensions and in the characteristics that are being measured.

The dielectric properties of wood are measured by using different techniques in accordance with the chosen frequency range. At low and at high frequencies, measurements with bridges are most widely accepted, while measurements at super high frequencies are carried out using waveguide and resonator techniques.

Below is a brief description of the methods used in measuring dielectric parameters of materials that have been employed for the determination of the dielectric characteristics of wood and wood-based materials. Every investigator whose works are mentioned or analyzed in this book used one or several of these methods in their experiments.

3.1 Measurement of the Dielectric Constant and Loss Tangent at Low Frequencies

For measurements of the dielectric properties of wood at frequencies up to 100 Hz, the bridge measurement method is used (Kasarnovskiy and Tareev 1980). The measurement of the wood DC entails the determination of the capacitance of the air capacitor introduced into the bridge arm with subsequent similar measurements carried out with the capacitor filled with the material to be tested. The ratio of these capacitance values determines the relative DC of the material. The compensation of the dielectric losses of the capacitor with the sample by the adjacent bridge arm variables determines the loss tangent of the wood.

Measurements of the capacitance C_x and $tg\delta$ can be performed with the help of the bridge, the schematic diagram of which is shown in Fig. 3.1. The bridge is energized through a step-up transformer.The voltage is applied between the bridge apex and the ground terminal. The two bridge arms adjacent to the "high voltage" point of the bridge comprise the capacitor C_x which is being tested and a high voltage capacitor C_0. The guarding electrode is connected to the grounded screen, and the high-voltage electrode is connected to the bridge apex. A variable resistor R_3 and a fixed resistor R_4 are cut into the two other arms. R_4 is shunted by a variable capacitor C_4. As an equilibrium indicator, a

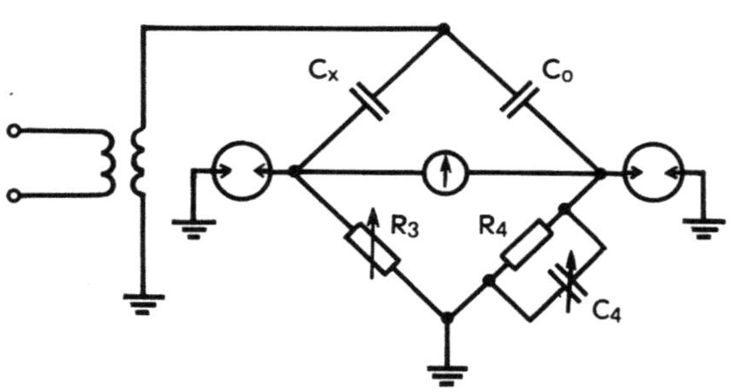

Fig. 3.1. A high-voltage bridge for measuring C_x and $tg\delta$: C_x capacitance of the sample; C_0 high-voltage capacitor; C_4 variable capacitor; R_3 variable resistor; R_4 fixed resistor

vibration galvanometer is used. Spark gaps are provided to protect the circuit in case of breakdown of the sample. The bridge is balanced by changing the resistance value of R_3 and the capacitance value of C_4. The bridge equilibrium is achieved when the products of impedances of the opposite arms of the bridge are equal to each other

$$Z_x \cdot Z_4 = Z_0 \cdot Z_3 \,. \tag{3.1}$$

In order to read the indication directly, the scale of R_3 should be calibrated in capacity units, and the scale of C_4 should be calibrated in the units of $tg\delta$. The dielectric loss tangent can be read directly from the instrument scale. When the instrument is not calibrated sufficiently, the value of $tg\delta$ can be calculated. The following equation can be used for the calculation of the wood loss tangent value

$$tg\delta = \omega R_4 C_4 \,. \tag{3.2}$$

It is necessary to measure or to calculate the capacitance value of C_0 of the unit without the sample while preserving the geometrical dimensions. The dielectric constant is determined by calculations using the results of the determination of the capacitance C_x

$$C_x = (C_0 \cdot R_4/R_3)/(1 + tg^2\delta) \,, \tag{3.3}$$

C_x and C_0 being known, the DC is

$$\varepsilon' = C_x/C_0 \,. \tag{3.4}$$

The measurements are carried out with the use of the three-electrode system which comprises the high voltage, the measuring, and the guarding electrodes (Fig. 3.2). As a material for the electrodes one can employ tin, lead, or aluminum foil, 10–50 μm thick. The foil is superimposed on the sample covered with silver paste and thoroughly rubbed to the sample surface to obtain a tight fit. A better method to achieve proper contact between the foil and the sample is described in the chapter dealing with the sample preparation process.

Measuring devices for determining the dielectric parameters of materials at low frequencies are now being produced industrially. For the purpose of carrying out measurements at different temperatures, the industry manufactures special measuring chambers securing the specified stable temperature with deviations not exceeding $\pm 1\,^\circ$C. Such chamber designs have been described (Ishida et al. 1959; Tsutsumi 1967; James 1975).

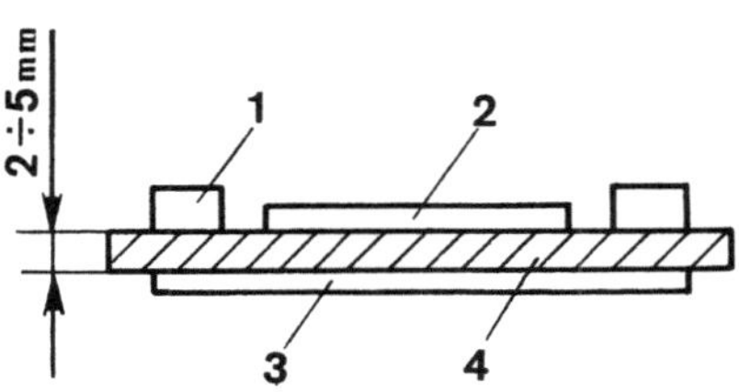

Fig. 3.2. The system of three electrodes and the sample: *1* guarding electrode; *2* measuring electrode; *3* potential-carrying electrode; *4* wood sample

3.2 Measurements of the Dielectric Constant and Loss Tangent at High Frequencies

For measuring dielectric charactetistics at frequencies above 100 Hz, the bridge or the resonance methods are used. There are some particular features in the measurement process in this frequency range. These features are associated with an increase of the influence of the edge effect, with the sample capacitance relative to the ground, and with the inductance and capacitance of the feeding wires. The methods used for measurement purposes provide compensation possibilities for eliminating the influence of the circuit stray inductance and capacitance. Such an elimination can be achieved by double and triple balancing of the bridge or by the use of special units with the two-electrode system and of the two-electrode units with an additional movable electrode.

The group of the capacitance type bridges is one of those groups used in practice. This group includes four-arm bridges that contain only active and capacitance elements. The circuit with the variable active resistance and capacitance is presented on Fig. 3.3. This circuit comprises two identical non-reactive resistors R_1 and R_2 in the bridge arms, a fixed resistor R_4, a variable resistor R_3, and changeable capacitors C_3 and C_4. The sample of capacity C_x is connected in parallel with the capacitor C_4. The bridge balance is adjusted by the change of values of C_4 and R_3, while endeavoring to keep the capacitance value of C_4 as low-as possible.

At a capacitance value equal to C'_4 and a resistance value equal to R'_3, the bridge can be considered balanced. For the second time, the bridge is balanced without the sample; the capacitance and the resistance values in this case will be C''_4 and R''_3, respectively.

The loss tangent is to be found from the following equàtion (Kasarnovskiy and Tareev 1980)

$$tg\delta = \frac{1}{\omega C_x R_x} = \frac{\omega (C''_4)^2 (R'_3 - R''_3) \cdot 10^{-12}}{C''_4 - C'_4}, \tag{3.5}$$

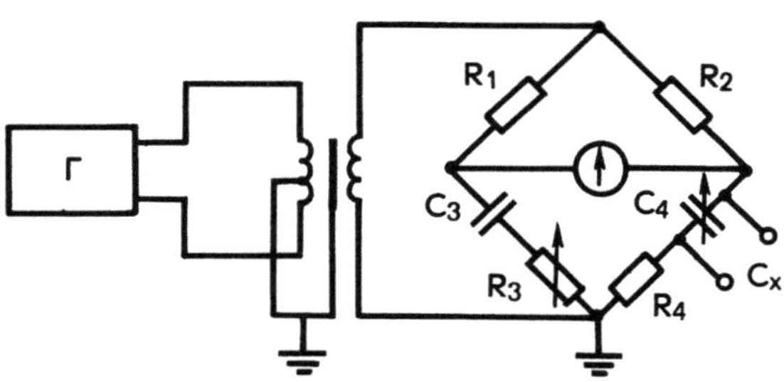

Fig. 3.3. The capacitance bridge with variable C and R values: C_3, C_4 capacitors; C_x capacitance of the sample; R_1, R_2, R_3, R_4 resistors; Γ generator

where

C'_4 and C''_4 are expressed in picofarads,

$$C_x = C''_4 - C'_4 \, . \tag{3.6}$$

Some requirements to which the circuit elements should correspond are as follows. The generator should ensure a stable frequency. The input transformer should have a range of values transformation ratio from 4 to 10 and be shielded from the ground. The resistors R_1 and R_2 should be equal to each other, the value being 5000 ohms. The variable capacitor C_4 should give a capacitance change from 50 to 1500 pF; its $tg\delta$ value should not exceed 10^4. Changes and readings of the capacitance value should be carried out with the help of a vernier, and the error of measurement should be within $\pm$ 0.2 pF. A decade resistance box of 1–10 000 ohms range should be used as the variable resistor. All elements of the bridge should be reliably screened.

For the purposes of the measurement of capacitances and dielectric loss tangent values, the following appliances are also used: bridge circuits with fixed resistors, bridge circuits with constant capacitance values, and transformer-type bridge circuits. Doubled transformer-type bridge curcuits are widely used nowadays.

The resonance circuits with lumped parameters are used in the frequency range from 10^3 to 2×10^8 Hz. Physical phenomenona in the resonance circuits are used for measuring the DC and loss tangent values. The resonance circuit methods are based on the applicaton of one high frequency generator with its tuned circuit into which the tested capacitor with the sample is introduced.

The resonance circuit method of susceptance variations is also widespread. The change of the susceptance is achieved by variations of the resonance circuit capacitance. The circuit includes a generator with a fixed frequency. The measuring resonant circuit is lightly coupled to the generator. The measuring circuit consists of an inductance coil and a variable capacitor with the sample connected to the capacitor in parallel. The generator operates in the invariable current regime, which results in corresponding changes in the voltage on the resonant circuit connected in parallel, namely, with the change of susceptance, the voltage value passes its maximum and then diminishes. The point of the highest voltage value developed by the resonant circuit corresponds to the resonance state of the circuit. Because of certain losses that are present in the circuit, the equivalent diagram includes besides L and C the conductivity g_k whose value corresponds to the value of the losses (Fig. 3.4a).

At the beginning, the circuit is tuned to resonance without the sample. The value of C_1 corresponding to the resonance state of the circuit is noted and the highest value of voltage U' developed by the circuit is determined. Changing the capacitance value in both directions from the circuit resonance point, the value of ΔC_1 can be determined, which corresponds to a tuning to the point where voltage U' diminishes to the value $U'/\sqrt{2}$. After the introduction of the sample into the circuit (Fig. 3.4b), the latter is tuned to resonance once more, and

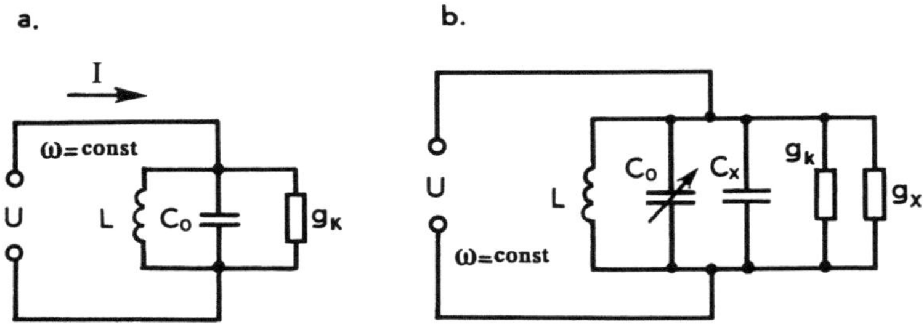

Fig. 3.4a, b. Equivalent circuits without **a** and with the sample **b**: C_0 capacitance of the calibrated capacitor; C_x capacitance of the sample; g_k and g_x conductances; L inductance

the new values are determined, namely the capacitance C_2, the circuit voltage U'', and the ΔC_2 value which corresponds to $U''/\sqrt{2}$. At the circuit resonance point, the inductive susceptance is equal to the capacitive susceptance, and therefore the admittance (full or total conductivity) of the circuit comprises only the real part of admittance $g_k = \omega \Delta C_1$ (Kasarnovskiy and Tareev 1980). The tuned circuit Q factor is determined as $1/Q_1 = g_k/\omega C_1$ or $1/Q_1 = \Delta C_1/C_1$

where

ΔC_1 = the change of capacity corresponding to the voltage U' diminishing to 0.707 U',

C_1 = the capacitance value at the circuit resonance state without the sample.

The circuit Q_2 factor (with the sample) is determined as $1/Q_2 = \Delta C_2/C_2$,

where

ΔC_2 = the change of capacity value corresponding to the voltage U'' diminishing to 0.707 U'',

C_2 = capacitance value with the sample.

With unchanged frequency, the circuit resonance state at the second tuning will occur when $C_2 + C_x$ is equal to C_1; hence, $C_x = C_1 - C_2$.

The loss tangent can be derived from

$$tg\delta = C_1(1/Q_2 - 1/Q_1)/C_x \tag{3.7}$$

and the DC of the sample

$$\varepsilon' = C_2/C_1 \; . \tag{3.8}$$

During the measurements, the voltage in the measuring network should be sine-shaped and of constant amplitude. The amplitude oscillation should not exceed $\pm$ 3%. The basic error of the frequency set-up should be within $\pm$ 1%.

A variation of the circuit resonance method is the method of determination of the dielectric characteristics by frequency variations. This method can also be used for determining the dielectric parameters of wood. The samples are put into special measuring units. The diameter of the electrode bearing the potential should not be less than the guarding electrode diameter. The width of the guarding electrode should not be less than double the thickness of the sample. The gap between the measuring and the guarding electrodes should not exceed 2 mm.

The two-electrode system is recommended for use in case the instruments do not provide possibilities to connect the guarding electrode and if the surface conductance can be neglected. Measurements at frequencies up to and including 1 MHz can be carried out with the electrode appliances (wiring system for connecting the sample with the instrument), which may be lever, screw, or spring-type devices, but which should meet the following requirements: the resistance of the feeding wires and their contacts should not exceed 0.03 ohms at d.c.; the pressure of the electrode on the sample should be $10 \pm 2\ \mathrm{kN/m^2}$.

For measurements at frequencies above 0.1 MHz, a measuring unit with a micrometric screw is recommended (Fig. 3.5) or a similar device. At frequencies exceeding 1 MHz, use of this kind of unit is obligatory. During the measurement process, the sample is set between the electrodes of the unit, while the movable electrode is lowered until the sample is gripped between the plates. The distance between the electrodes is to be read from the micrometer head. If the electrodes are attached to the sample, their thickness should be subtracted from this reading. For the second set of measurements, the sample should be taken out of the unit, and the distance between the electrodes should be set so that the capacitance value of the measuring unit is the same as it was for the first measurements (with the wood sample installed in the unit).

In case the micrometric electrodes are used for the contact method of measurements, the dielectric constant value of the samples can be determined from the equation

$$\varepsilon' = a'/a'' \tag{3.9}$$

where

a' = thickness of the sample
a'' = distance between the capacitor plates without the sample.

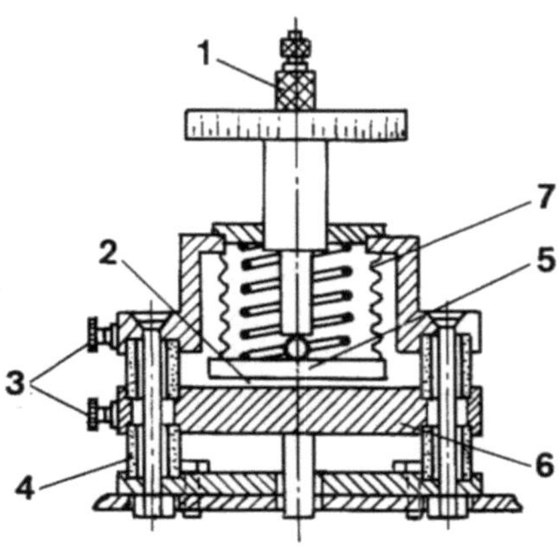

Fig. 3.5. The measuring appliance with a micrometer screw: *1* micrometer head; *2* space for the sample; *3* terminals; *4* fused quartz; *5* movable electrode; *6* fixed electrode; *7* metal corrugated tube

When using measuring instruments and installations accommodated to direct measurements of the dielectric loss tangent, the value of $tg\delta$ for the sample can be found from the equation

$$tg\delta = \frac{C_1}{C_x}(tg\delta_1 - tg\delta_2) , \tag{3.10}$$

where

$tg\delta_1$ and $tg\delta_2$ = loss tangent values of the capacitor with and without the sample

C_1 = total capacitance of the capacitor with the sample in it (pF)

C_x = the capacitance of the sample determined from the following equation:

$$C_x = 0.0695 d^2/a'' , \tag{3.11}$$

where

d = the electrode diameter.

When measuring the dielectric characteristics with the instruments and installations adjusted for determination of the Q factor of the tested circuit by the resonance methods, the loss tangent can be found from the equation

$$tg\delta = \frac{Q_1 - Q_2}{Q_1 \cdot Q_2} \cdot \frac{C_p}{C_x} , \tag{3.12}$$

where

Q_1 and Q_2 = Q factors of the circuit without and with the sample inserted between the capacitor plates

C_p = total capacitance of the circuit in pF, equal to the standard capacitor of the tuned instrument when the measured capacitor is disconnected

C_x = capacitance value of the sample being tested which is derived from Eq. (3.11).

If it is necessary to measure the dielectric characteristics at different temperatures, the measuring unit should be placed in a chamber furnished with temperature control.

3.3 Measurement of the Dielectric Constant and Loss Tangent at Super High Frequencies

The rise in frequency in instruments that use circuits with lumped constants and operate with the use of the susceptance variation principle results in a reduction in the circuit Q factor because of the skin effect influence and radiation losses in the induction coil. At frequencies above 300 MHz, the employment of circuits with lumped parameters causes a considerable increase of errors which requires correction of the results. This is the reason why at frequency ranges above 3×10^8 Hz circuits with distributed constants should be used.

For the purpose of measuring complex DCs of wood, the resonance or waveguide methods may be used. The principal possibilities of both methods are identical. However, the resonance method has a number of faults. It requires a resonator with a high quality factor of its own ($Q \geqslant 5000$ at frequency 10 GHz); dependence of the sample diameter on the losses in the sample; the necessity of measuring the frequency difference with an accuracy of ± 1 MHz, which requires the use of wavemeters with a high frequency resolution.

The waveguide method is free of these faults, its performance being comparatively simple, and the computer-assisted data processing permits one to obtain ε' and $tg\delta$ values quickly and with sufficient accuracy.

As recommended by von Hippel (1954), to determine the complex DC, the waveguide short-circuit method was used. The essence of the method lies in the evaluation of the input impedance of the line with the dielectric sample Z_{bx} (Fig. 3.6) expressed through the wave propagation constant in the sample

$$\gamma_1 = \alpha + i\beta \,, \tag{3.13}$$

where

α = attenuation constant
β = phase constant.

The propagation constant γ_1 in its turn is determined by the complex DC

$$\gamma_1 = (2\pi/\lambda_0)\sqrt{(\lambda_0/\lambda_{kp})^2 - (\varepsilon' - i\varepsilon'')} \,, \tag{3.14}$$

where

λ_0 = wavelength in free space
λ_{kp} = critical wavelength in the line.

The input impedance of the short-circuited line without losses is determined from the equation

$$Z_{bx} = iZ'_0 tg\beta l = Z'_0 thi\beta l \,, \tag{3.15}$$

where

$tg\beta$ = tangent of β
$thi\beta$ = hyperbolic tangent of $i\beta$
Z'_0 = characteristic impedance or equivalent resistance of the line filled with dielectric
l = length of the line filled with dielectric (the sample length).

As the losses in the dielectric are not equal to zero, the factor $i\beta$ in Eq. (3.15) should be replaced with the propagation constant

$$\gamma_1 = \alpha + i\beta \,.$$

Then

$$Z_{bx} = Z'_0 th\gamma_1 l \,, \tag{3.16}$$

where

$th\gamma_1$ = hyperbolic tangent of γ_1.

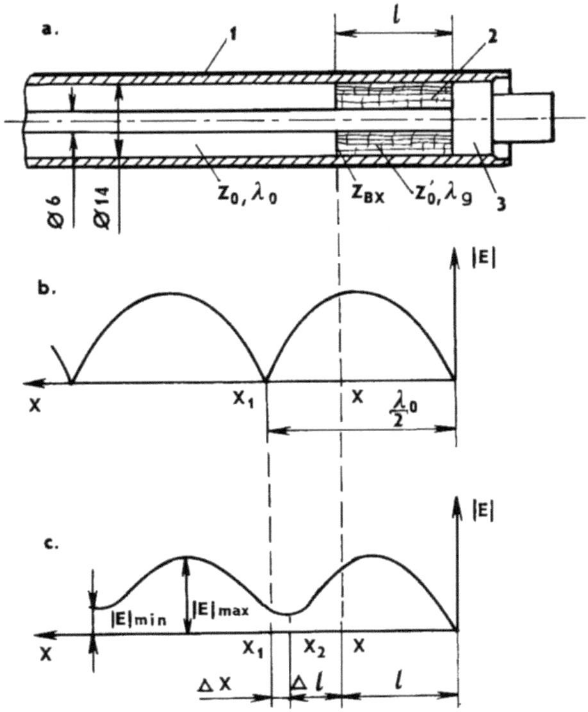

Fig. 3.6a–c. Graphic illustration to the waveguide method for measurement of ε' and $tg\,\delta$ values. **a** Wood sample position in the coaxial short-circuited line; **b** distribution of electric field E value along the line without the sample; **c** distribution of electric field E along the line with the sample placed at the end of the line. *1* Coaxial line; *2* the sample; *3* the short-circuiting piston; l the sample length; Δl distance from the sample end face surface to the first minimum value of E; x, x_1, x_2 coordinates; Δx the bias value of the E minimum point; Z_0 characteristic resistance of the line filled with air; Z'_0 characteristic resistance of the line with the positioned sample; Z_{bx} input resistance of the line with the positioned sample; λ_0 and λ_g wavelength values in the line and in the sample

Expressing Z'_0 through the characteristic impedance Z_0 of the line filled with air, we find that

$$(1/i\beta_0 l)/(Z_{bx}/Z_0) = th\gamma_1 l/\gamma_1 l\,, \tag{3.17}$$

where

$\beta_0 = 2\pi/\lambda_b$ is the phase constant of the waves in the line filled with air.

As the values of the travelling-wave ratio K_r and the phase of the standing wave θ are simple to measure, Eq. (3.17) can be reduced to the following form

$$\frac{th\gamma_1 l}{\gamma_1 l} = -i\frac{\lambda_0}{2\pi l}\cdot\frac{K_r - itg\theta}{1 - iK_r tg\theta}\,, \tag{3.18}$$

where

K_r = traveling-wave ratio, $K_r = E_{min}/E_{max}$
E_{min} and E_{max} = voltage minimum and maximum in the line
θ = phase of the standing wave.

In case use is made of a coaxial line in which vibrations of the TEM-type (the transverse electromagnetic mode has electric and magnetic fields transverse to the direction of energy flow) are produced, the wavelength in the line coincides with the wavelength in free space, $\lambda_0 = \lambda_b$, $\lambda_{kp} = \infty$, and in Eq. (3.18) the propagation constant takes the following form

$$\gamma_1 = (2\pi/\lambda_0)\sqrt{-\varepsilon' + i\varepsilon''} . \tag{3.19}$$

In Eq. (3. 18), the values of l and λ_0 are specified, while K_r and θ are the values to be measured. Solving Eq. (3.18) relative to γ_1, the values of ε' and $tg\delta$ can be derived from the equations

$$\varepsilon' = (\beta^2 - \alpha^2)/\beta_0^2 , \tag{3.20}$$

$$tg\delta = 2\alpha\beta/(\beta^2 - \alpha^2) , \tag{3.21}$$

where

β_0 = the phase constant of the wave in free space.

Figure 3.7 represents a schematic diagram of the measuring instrument. The device comprises a generator producing standard signals of the required frequency range, a ferrite valve or attenuator, a measuring line, a microammeter, and a section of coaxial (or waveguiding) line with the piston short-circuiting the line and a casing. Around the coaxial line, a thermally insulated casing is to be fixed. The casing should be provided with inlet and outlet branch pipes for feeding it with warmed liquid when measurements at various positive temperatures are required.

For measurements at negative temperatures, a thin-walled copper pipe is wound around the coaxial tube in which the sample is placed. Through the copper pipe pass nitrogen vapors from a Dewar flask. Cooling of the sample to temperatures down to − 40 °C can thus be achieved.

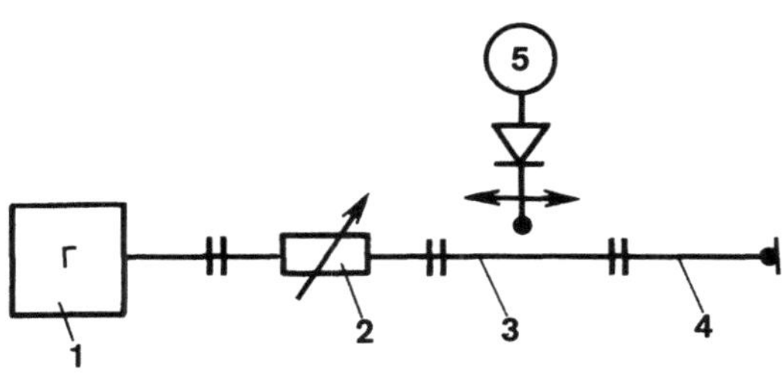

Fig. 3.7. The measuring instrument with the coaxial measuring section: *1* generator; *2* discriminating attenuator or ferrite valve; *3* measuring line; *4* the section of the coaxial line; *5* microammeter

The thoroughly prepared sample (Sect. 3.4) in the form of a bushing with the outer and inner diameters equal to 14 and 6 mm, respectively, is put on the central conductor and tightly pressed with the piston. From experience, it was estimated that for measurements at frequencies from 1 to 6 GHz, the sample length should be 20–40 mm.

Before starting measurements, the generator must be tuned to the specified frequency and SHF waves passed through the coaxial cable and ferrite valve to the measuring coaxial line and the measuring waveguide. The ferrite valve is a device that lets the super high frequency waves pass in the "straight" direction and does not let them go "backwards", absorbing the reflected energy. The wave reflected from the short-circuited piston and the incident wave form standing waves. The carriage with the measuring line probe (which is connected to the microammeter through a rectifier) is moved until the node of the standing wave nearest to the waveguide is found. The position of the node should be estimated with an error not exceeding 0.05 mm. At the same time, the wavelength should be checked by measuring the distance between the nodes by the so-called "fork" method. This makes it possible to find the position of the minimum reading with higher accuracy. According to the "fork" method, the two positions of the measuring line probe should be noted on both sides of the point of minimum value where the indicator (microammeter) readings are the same. The midpoint between these positions is assumed to be the minimum. The distance between these two positions of the probe should not exceed 0.02–0.05 of the λ_0 value.

The measurement technique amounts to performing the following operations (Fig. 3.6):

1. With the help of the measuring line, the position of any minimum value of the electric field strength E in the short-circuited line without the sample can be found, e.g., point x_1 (Fig. 3.6b).

2. Similarly, the point of minimum value of the standing wave with a sample l long can be found, e.g., point x_2 (Fig. 3.6c). The standing wave minimum value, which is not equal to zero because of the losses in the wood, shifts in the direction of the sample as the wavelength λ_g in the sample becomes less than λ_o.

3. The shift of the standing wave minimum point is determined from

$$\Delta x = x_1 - x_2 .$$

4. The phase of the standing wave θ in the plane x determined by the distance Δl is to be calculated as follows:

$$\Delta l = \frac{\lambda_0}{2} - l - \Delta x$$

$$\theta = 2\pi \Delta l / \lambda_0 .$$

5. The traveling-wave ratio is determined

$$K_r = \frac{E_{min}}{E_{max}} .$$

In case the current voltage characteristic of the measuring line crystal detector has a square-law form, the value of K_r will be

$$K_r = \sqrt{I_{min}/I_{max}} \, ,$$

where

I_{min} and I_{max} = minimum and maximum readings from the microammeter.

If $K_r \leqslant 0.1$, the traveling-wave ratio is

$$K_r = \pi \Delta x'/\lambda_0 \, ,$$

where

$\Delta x'$ = distance between the two positions of the measuring line probe corresponding to the doubled readings from the microammeter relative to its minimum reading value.

Equation (3.18) can be solved with the help of a computer. The program has to be prepared so that the specified values are Δx, $\Delta x'$, λ_0, and l, while the unknown quantities are ε' and $tg\delta$.

It is worthwhile to note that Eq. (3.18) has several solutions of which only one is to be chosen. The selection of the required values of ε' and $tg\delta$ is not difficult as the adjacent values of the DC and loss tangent differ several times from the quantities to be determined.

In the coaxial waveguide, the waves of type TEM are so propagated that the electric field is always radically directed relative to the waveguide. For cylindrically shaped samples with the axis parallel to the fiber direction L (Fig. 3.8a), the electric field strength vector is transverse to the fiber direction simultaneously in radial and tangential directions and at an angle to these directions. That is why the measured values of dielectric properties are actually the mean values of the indexes obtained for the radial and tangential directions. Similarly, when the

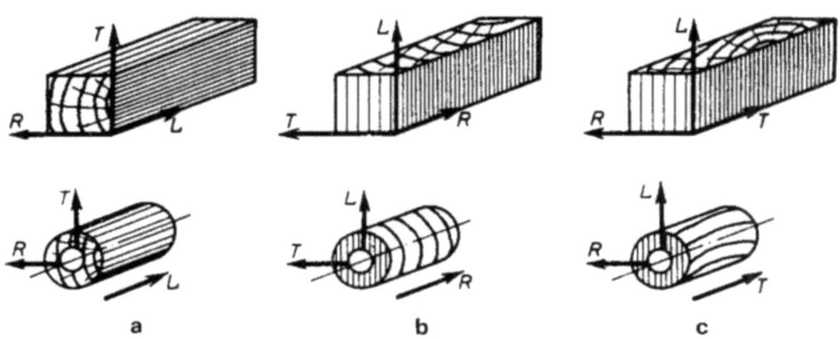

Fig. 3.8a–c. Lumber blanks and the samples intended for measurement of the dielectric properties at super high frequencies using a coaxial line. The sample axis corresponds to the directions: **a** longitudinal; **b** radial; **c** tangential; L, R, T longitudinal, radial, and tangential directions of the wood axes of anisotropy

axes of the samples are parallel with the radial direction relative to the growth rings (Fig. 3.8b), the measured parameters are the mean values of ε'_{TL} and $tg\delta_{TL}$ in the tangential and longitudinal directions. When the sample axes are tangentially oriented (Fig. 3.8c), the measured parameters are the mean values of ε'_{RL} and $tg\delta_{RL}$ in the radial and longitudinal directions.

In order to calculate the wood parameters in every direction (R, T, and L), it is assumed that the values obtained are the simple mean parameters in two directions. Then the system of three equations with three unknown values ε'_R, ε'_T, and ε'_L are formed

$$\begin{cases} \varepsilon'_R + \varepsilon'_T = 2\varepsilon'_{RT} \\ \varepsilon'_R + \varepsilon'_L = 2\varepsilon'_{RL} \\ \varepsilon'_T + \varepsilon'_L = 2\varepsilon'_{TL}. \end{cases} \tag{3.22}$$

Since the value of ε'_{RT}, ε'_{RL}, and ε'_{TL} are known, the values of ε'_R, ε'_T, and ε'_L can easily be calculated. A similar system of equations is to be worked out for the determination of the loss tangent values.

It was established experimentally that the values of ε'_{RL} and ε'_{TL} are very near each other, and thus the conclusion may be drawn that $\varepsilon'_R = \varepsilon'_T = \varepsilon'_\perp$ and that $\varepsilon'_L = 2\varepsilon'_{TL} - \varepsilon'_\perp$ or $\varepsilon'_L = 2\varepsilon'_{RL} - \varepsilon'_\perp$. The same reasoning can be applied to $tg\delta$.

Measurement of the dielectric properties of wood with the use of a coaxial line is valid at a frequency range up to 5–6 GHz. When the distance between the coaxial electrodes comes near to the wavelength, measurements are carried out with the use of hollow round or rectangular waveguides (James and Hamill 1965; Peyskens et al. 1984). Measurement of the wood DC and $tg\delta$ at frequencies from 6 to 10 GHz can be done with the instrumentation shown in Fig. 3.9. The measuring section of the waveguide for determination of the dielectric parameters at a frequency of 10 GHz is a rectangular tube with a cross-section equal

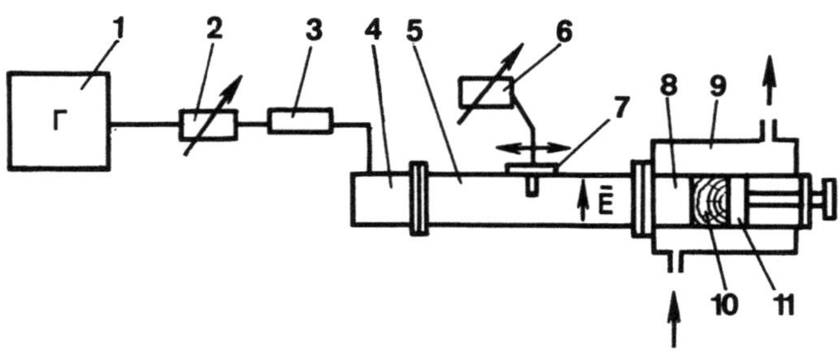

Fig. 3.9. The measuring instrumentation with a rectangular measuring section of the waveguide: *1* generator; *2* attenuator; *3* tuner; *4* coaxial waveguide transition; *5* waveguide measuring line; *6* device for measuring the travelling wave factor value; *7* sliding probe; *8* measuring section of the waveguide; *9* casing for the flowing liquid; *10* the wood sample; *11* short-circuiting piston; $\overline{E}$ the vector of electric field strength

to 23 × 10 mm, which is furnished with the sliding short-circuiting piston. The micrometer head serves for precise positioning of the piston. The waveguide is covered with a heat-insulating casing, which is fed with liquid of a given temperature from a thermostat. The sample should be of the same cross-sectional dimensions as the waveguide and should sit tightly and without gaps in the latter. Before starting the experiment, the sample should be inserted into the waveguide so that it closely adjoins the short-circuiting piston. The waveguide flange should be connected to the measuring line.

In the waveguide without the sample, a standing wave is formed. The distance between the nodal points of the wave is equal to half of the wavelength in the waveguide $\lambda_b/2$. The electric field strength in the nodes is equal to zero, as the amplitudes of the forward and the reflected waves are equal to each other. The picture changes when the sample is inserted into the device, as the superposition of three waves takes place, namely, the incident wave and the two reflected waves, one of which is reflected from the sample, while another comes from the short-circuiting piston. As a result of this superposition, the field strength at the point of the minimum value of the resultant wave is not equal to zero. In addition, the point of minimum value is shifted toward the sample.

The distance from the end face of the sample to the nearest point of minimum, the travelling-wave factor, and the propagation constant are mutually related in an equation similar to Eq. (3.18) in which the wavelength in the free space is replaced by the wavelength in the waveguide

$$\frac{th\gamma_1 l}{\gamma_1 l} = -i\frac{\lambda_b}{2\pi l}\cdot\frac{K_r - itg\theta}{1 - iK_r tg\theta}, \tag{3.23}$$

where

$\lambda_b =$ wavelength in the waveguide which is equal to

$$\lambda_b = \lambda_0/\sqrt{1-(\lambda_0/\lambda_{kp})^2}\,, \tag{3.24}$$

$\gamma_1 =$ propagation constant in the sample, which in this case is bound with the complex DC of the sample through the following relation

$$\gamma_1 = (2\pi/\lambda_0)\sqrt{(\lambda_0/\lambda_{kp})^2 - \varepsilon' + i\varepsilon''}\,. \tag{3.25}$$

The critical wavelength in the waveguide λ_{kp} in the general case is determined from

$$\lambda_{kp} = 2/\sqrt{(m/a)^2 + (n/b)^2}\,, \tag{3.26}$$

where

m and $n =$ figures which identify the type of oscillations in the waveguide
a and $b =$ dimensions of the narrow and the wide walls of the waveguide.

In the case of wave type H_{10} for which $m = 1$ and $n = 0$, the critical wavelength is $\lambda_{kp} = 2a$.

The remaining values of Eq. (3.23) are the same as in Eq. (3.18).

After solving Eq. (3.23) relative to the propagation constant γ_1, the values of ε' and $tg\delta$ can be found from the equations

$$\varepsilon' = [(2\pi/\lambda_{kp})^2 + \beta^2 - \alpha^2]/\beta_0^2 \,, \tag{3.27}$$

$$tg\delta = 2\alpha\beta/[(2\pi/\lambda_{kp})^2 + \beta^2 - \alpha^2] \,. \tag{3.28}$$

For the determination of the dielectric parameters of the sample, it is necessary to find the values of Δx, $\Delta x'$, λ_b, and l. After these values have been found, it is possible to determine the values of ε' and $tg\delta$ of the sample using Eqs. (3.23–3.28). The calculations can be carried out with a computer.

In the rectangular type waveguide with the wave type H_{10}, the electric field is always directed parallel to the narrow wall (Fig. 3.9). Therefore, by changing the orientation of the sample in the waveguide, it is possible to measure the dielectric characteristics of wood independently in any of the radial, tangential, and longitudinal directions.

It should be noted that at frequencies of up to 3–4 GHz the dielectric characteristics of the samples with any moisture content can be measured by means of the methods described. At higher frequencies, the sample with a high moisture content can short-circuit the waveguide, and the incident wave will not penetrate into the sample and will be completely reflected from the sample end face surface. This is the reason why the waveguide short-circuit method at frequencies above 3–4 GHz is only valid for measurements of samples with a moisture content up to 19–25%.

An estimation of systematical errors of the waveguide method of measuring ε' and $tg\delta$ is presented by von Hippel (1954). The error in measuring ε' depends upon the error of measuring the distance from the sample to the first minimum point of the standing wave Δl (Fig. 3.6), on the goodness of fit of the samples and conductors, on the error of measuring the sample length, and on the travelling-wave factor.

Measuring lines with the scale division equal to 0.02–0.05 mm makes it possible to measure the Δl distance with an error of about 0.5%. The error in measuring the traveling-wave factor is about 2%. The error caused by the tightness of the sample fit is 0.5%, and that of measuring the sample length is 1–1.5%. The maximum value of the systematic error of the DC measurements estimated as the mean square of the errors enumerated above is equal to 2.5–2.8%.

The systematic error of the measurement of $tg\delta$ in the waveguides mainly depends on the error in the measurement of the traveling-wave factor, which is about 2%, and on the losses in the waveguide, which are 1–1.5%. The total relative systematic error of the $tg\delta$ measurements is estimated at 2.5–3.0%.

Casual errors are caused by distortions of the geometric form of the samples, the wood structure variability, moisture content variations, or other factors. These errors exceed by several times the value of the systematic errors, especially when moist samples are being tested. Therefore, as a degree of the total relative error, the larger value should be taken into account, namely the casual error value.

For measuring the dielectric characteristics of wood at super high frequencies, the resonator method can be used as well (Brandt 1963). However, I and practically all research workers who have studied the dielectric properties of wood and wood-based materials have used the waveguide method described above.

3.4 Preparation of Samples

Selection of the samples for measurement of the dielectric characteristics of wood as a rule is carried out in two steps. Initially, bars are cut (Fig. 3.8) from the boards sawed out of the inner wood. The boards should be of the radial and the tangential sawing. The cross-section of the bars should exceed the dimensions of the future samples by 10–15 mm. The bars with the fewest blemishes are humified to a moisture content approaching the required percentage, and samples of the specified dimensions are prepared. The number of samples should exceed the quantity required for measuring purposes by 20–50%. The surplus is intended for possible rejection of faulty samples revealed during the preparation process. Measurement of the dimensions of the finished samples should be made with an error not exceeding $\pm$ 0.1 mm. The roughness of the sample surface, which is characterized by the mean value of the maximum heights of the surface unevenness, should be no more than 60 μm.

The natural variability of wood and the changes of conditions in which the measurements of different samples take place considerably affect the data obtained. This is why it is necessary to test several samples to obtain reliable results. The measured data are processed by an analysis of variance method. In the process, the following statistical characteristics are determined: arithmetic mean, root mean square deviation, coefficient of variation, average error of arithmetical mean value, and accuracy index.

During the experimental study of the dielectric properties, it is necessary to take into account the macrostructure of wood. This means that the samples should contain a sufficient quantity of the wood macrostructural elements (earlywood and latewood, rays, vessels). Theoretical analysis shows that the elementary volume of wood that contains a sufficient quantity of all its macrostructural heterogenities should not be less than 1 cm^3. Therefore, the samples should have a volume not less than the elementary volume and should exceed 1 cm^3.

In a study of wood properties the accuracy index of the mean values determination should not exceed 5%. During the DC measurements, this requirement is practically always fulfilled. But because of the wide range of variations of the $tg\delta$ values in wood with a high moisture content, the accuracy index for the loss tangent measurements can be up to 6–8%.

Our research showed that the coefficient of variation of the DC value increases with an increase of moisture content in the wood from 2 to 14%, while that of the $tg\delta$ changes from 2 to 27%. Usually, for obtaining the above-mentioned accuracy indexes, it is necessary to measure 6–20 samples for each fixed value of moisture content and temperature.

Most often, measurements at low and high frequencies are carried out with samples made in the form of disks with a diameter of 35–80 mm and a thickness of 2–5 mm. Rectangular samples are also acceptable. However, round samples are easier to manufacture because the use of a wood-working machine makes it possible to obtain them with precise overall dimensions. While preparing the samples, it should be remembered that one of the principal axes of anisotropy (L, R, or T) should be parallel to the axis of the disk. The allowable deviation of the side surface of the sample to the end face surfaces should not exceed 0.1 mm.

The preparation of the moist samples is the most labor-consuming process. For obtaining surfaces with fine-grain roughness, the rated speed of the wood-working machine should be sufficiently high. However, machining the blank of a specified moisture content at high speeds causes warming-up and drying of the surface layers of the samples because of friction between the wood and the cutting tools. After machining, the samples must therefore be conditioned, i.e., humified again, which results in their enlargement. In order to finish the preparation and obtain the required dimensions of the samples, the procedure of machining and conditioning should be repeated several times.

In order to obtain results that can be reproduced in the subsequent experiments at low and high frequencies, it is necessary to exclude variations of electrical and chemical conditions on the surface of the samples. This aim can be achieved by covering the sample surfaces with electrically conductive silver paint. To prevent the loss of moisture, gluing aluminum foil to the surface of the samples with electrically conductive paste is recommended, after the surfaces have been covered with the silver paint. The end face surfaces of the samples should be jacketed with a thin plastic tape.

The waveguide method is the most convenient for measuring the dielectric characteristics at super high frequencies, but this method requires an extremely thorough preparation of the samples. The samples for testing in a coaxial line must be made in the form of bushings with external and inner diameters corresponding to the diameters of the conductors (Fig. 3.8). The length of the specimens should be 20–40 mm. The sample axes should be parallel with one of the principal axes of anisotropy (L, R, or T). The samples are prepared on a wood-turning lathe. At the beginning, the hole in the lumber blank should be drilled and ground. Then the blank should be put on the mandrel and its external and end face surfaces thoroughly turned and ground. Special efforts must be undertaken to secure the lowest possible parallel misalignment of the end faces of the sample.

The moisture content of the blank should be approximately equal to that of the finished sample. The treatment of the moist samples creates difficulties similar to those described above connected with drying-out the treated surfaces and subsequent changes of the sample dimensions during the adjustment of the moisture content process. This is why the finishing of the samples should be carried out two or three times after seasoning the samples to the required moistening condition.

It is necessary to note one more complication connected with the sample preparation. The fact is that the annual rings of earlywood and latewood are of differing density and have different mechanical properties. This is the reason for the appearance of unevenness on the surface of the moist samples, namely ridges on the latewood and recesses in the earlywood. In order to reduce these, it is recommended to turn the blanks at a low horizontal motion speed of the lathe carriage and to use sharp cutting tools, removing layers of 0.02 to 0.04 mm thickness per pass. The error of the length measurements should be within ± 0.1 mm.

Checking of the sample dimensions is carried on with a special template (Fig. 3.10) that is used as a measuring gauge as well. After the final conditioning, a fluffiness appears on the side surfaces of the sample. Because of this fluffiness, the samples must be calibrated before measurement by putting them on the inner part of the template and subsequently pressing them through the outer part of the template (Fig. 3.10). The sample should sit closely and tightly on the inner part of the template and should enter tightly into its outer part. Only after completion of the above operations is the sample considered ready for measurement and can be put into a hermetic chamber.

As the wood swells during the process of humidification from the oven-dry state to the fiber saturation point, it is necessary to make separate samples for measuring at each of the required moisture levels within the above moisture limits. The moisture content of the sample blanks should be 2–3% higher than that required for measurement purposes. If the wood moisture content should exceed the fiber saturation point, the same samples can be used repeatedly for measurements, and their moisture content can be varied as required.

For measurements of the wood dielectric properties in a waveguide of rectangular shape, the samples should also be rectangular and in conformity with the waveguide dimensions (usually, its cross-section is 10 × 23 mm). The length of the samples should be 5–30 mm. The preparation of such samples is

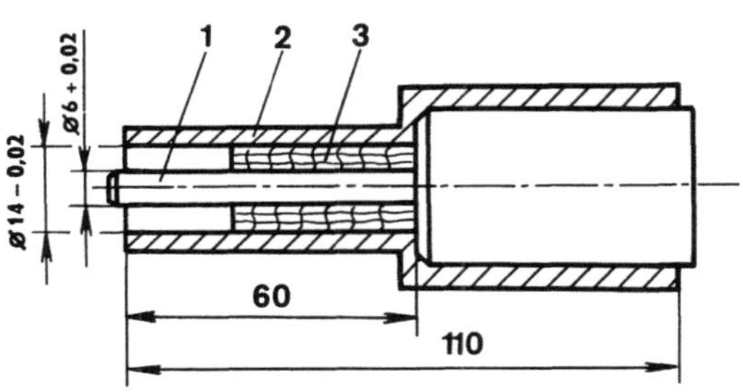

Fig. 3.10. Template for checking and calibrating the samples: *1* inner part of the template; *2* outer part; *3* the sample under test

a more labor-consuming process as the precise treatment of flat surfaces of moist wood is more complicated, and the dimension tolerances of the samples exceed those of samples made with a wood-turning lathe. Rectangular samples also require multiple finishings after moisture conditioning and calibration.

The temperature expansion of the samples can be neglected as the coefficient of linear expansion of wood is very small, namely from 3×10^{-6} to $40 \times 10^{-6}\,K^{-1}$. Measurements of the dielectric characteristics at temperatures from 40 to 90 °C require an inspection of moisture content of the samples before and immediately after the measurement has been made.

The methods of measuring the dielectric characteristics of wood described in this chapter can also be used for testing wood-based materials such as fiberboards, chipboards, modified wood, etc. Besides the methods described above other procedures can also be applied. These methods have been described by von Hippel (1954), Brandt (1963), and Bergmann (1969), for different kinds of dielectrics. The selection of a proper measurement technique mainly depends on the possibilities available for preparing the samples, bearing in mind that the samples should represent the macrostructure of wood. This means that the specimens should not be smaller than the elementary macrostructural volume (1 cm^3). It is necessary also to take into account that the dielectric parameters of dry and moist wood differ by hundreds and thousands of times on measurements at low frequencies and tenfold when the samples are tested at high frequencies. This is the reason why it is not always possible to encompass all required ranges of moisture content and of temperatures with a single frequency range measurement.

Chapter 4

Dielectric Properties of Oven-Dry Wood

4.1 Influence of Anisotropy, Density, and Frequency on the Dielectric Constant, Loss Tangent, and Loss Factor

The term oven-dry wood means wood dried to a constant mass in air at a temperature equal to $+103 \pm 2\,°C$. Research has shown that the dielectric parameters of oven-dry wood are considerably influenced by the wood density, temperature, the external field frequency, and the orientation of vector $\bar{E}$ relative to the direction of the fibers. In addition, wood from different tree species possesses different dielectric properties. An increase in the field strength affects the ε' and $tg\delta$ values of oven-dry wood inconspicuously (Nanassy 1972).

The dielectric parameters of oven-dry wood are geometrically interpreted by an ellipsoid, the axes of which are directed along the principal axes of anisotropy (Fig. 1.8), and if vector $\bar{E}$ is oriented in any arbitrary direction, the dielectric characteristics can be determined from Eqs. (1.32), (1.33), (1.36), and (1.37) if the values of ε' and $tg\delta$ in the directions of L, R, and T are known. This is the reason why studies of the dielectric properties of wood are carried out with the vector $\bar{E}$ oriented in these particular directions.

For the same wood density and field frequency, the largest ε' and $tg\delta$ values of oven-dry wood will be obtained when vector $\bar{E}$ is oriented parallel to the fibers. This can be explained by the structure of the wood. The elongated wood cells (tracheids and fibers) are oriented in the same direction as the stem, and the microfibrils in the cell walls are largely oriented in this direction. Cellulose chains, which form microfibrils, possess larger dielectric characteristics in the fiber direction. With other conditions unchanged, the following relations apply: $\varepsilon'_L > \varepsilon'_R > \varepsilon'_T$ and $tg\delta_L > tg\delta_R > tg\delta_T$. In the radial and tangential directions, the difference in dielectric parameters is less significant. The larger values of ε' and $tg\delta$ with vector $\bar{E}$ oriented in the radial direction relative to the tangential orientation can be explained by the presence of rays which are radially oriented. Assuming that the values of the dielectric parameters of the rays with the vector $\bar{E}$ orientation parallel to the rays are equal to the similar parameters of wood with the same orientation of $\bar{E}$ and also supposing that the influence of the rays on the dielectric parameters is proportional to their volume, the following equations can be stated

$$\varepsilon'_R = (1 - V_r)\,\varepsilon'_T + V_r \varepsilon'_L\,, \tag{4.1}$$

$$tg\delta_R = (1 - V_r)\,tg\delta_T + V_r tg\delta_L\,, \tag{4.2}$$

where

V_r = the relative volume of the rays ,

or

$$\varepsilon'_R/\varepsilon'_T = (1 - V_r) + V_r(\varepsilon'_L/\varepsilon'_T) , \tag{4.3}$$

$$tg\delta_R/tg\delta_T = (1 - V_r) + V_r(tg\delta_L/tg\delta_T) . \tag{4.4}$$

The relative volume of rays in most coniferous species is within 5–10%, that of such species as birch, aspen, elm, chestnut, and ash is 11–12% and is 22–28% in beech and oak. The relation $\varepsilon'_L/\varepsilon'_T$ and $tg\delta_L/tg\delta_T$ in the frequency range from 10^2 to 10^{11} Hz and at $t = 20\,°C$ for oven-dry wood lies within 1.1–2.1. Then the calculated relations of $k_{1(R/T)} = \varepsilon'_R/\varepsilon'_T$ and $k_{2(R/T)} = tg\delta_R/tg\delta_T$ for different wood species and at various frequencies change within the limits specified in Table 4.1.

The value of $k_{1(R/T)}$ at frequencies from 10^2 to 10^8 Hz averages 1.04 – 1.08 and at frequencies $10^9 - 10^{11}$ Hz, 1.02 – 1.04, while the value of $k_{2(R/T)}$ is equal to 1.06 – 1.19 and 1.04 – 1.10, respectively. The correctness of the relations obtained are confirmed by results reported by Kröner and Pungs (1953), Berdinskikh et al. (1955), Birjukov (1964), Rafalski (1966), and Norimoto (1976).

For most coniferous and angiosperm species, the difference between ε'_R and ε'_T is equal to 2–5% and that between $tg\delta_R$ and $tg\delta_T$, 4–9%. This difference is often not taken into account, and it is assumed that $\varepsilon'_R = \varepsilon'_T$ and $tg\delta_R = tg\delta_T$. For wood species made up of 18–28% rays, such an assumption entails an error of about 10–20%.

The literature usually give average values for dielectric parameters for cases in which the vector $\bar{E}$ is perpendicular to the fiber direction without a separation of the R and T directions. If it is necessary to take into account the difference of dielectric parameters in the R and T directions, the values of ε'_R and ε'_T ($tg\delta_R$ and $tg\delta_T$) can be found from the relations

$$\varepsilon'_R = \frac{2\varepsilon'_\perp}{1 + 1/k_{1(R/T)}}, \quad \varepsilon'_T = \frac{2\varepsilon'_\perp}{1 + k_{1(R/T)}} , \tag{4.5}$$

$$tg\delta_R = \frac{2tg\delta_\perp}{1 + 1/k_{2(R/T)}}, \quad tg\delta_T = \frac{2tg\delta_\perp}{1 + k_{2(R/T)}} , \tag{4.6}$$

where values of $k_{1(R/T)}$ and $k_{2(R/T)}$ can be taken from Table 4.1.

An example of a calculation of ε'_R and ε'_T values is as follows. Let us determine the DC value of oven-dry birch wood at an orientation of $\bar{E}$ in the radial and tangential directions under the following conditions: $\rho_0 = 0.6$ g/cm^3, frequency $f = 10^8$ Hz, temperature $t = 20\,°C$.

From Appendix 1 we find that $\varepsilon'_\perp = 2.1$; from Table 4.1 we take the average value of $k_{1(R/T)}$ which is equal to 1.04. Using Eq. (4.5), we determine that

$$\varepsilon'_R = \frac{2\varepsilon'_\perp}{1 + 1/k_{1(R/T)}} = \frac{2 \cdot 2.1}{1 + 1/1.04} = 2.14$$

Table 4.1. Relations $k_{1(R/T)} = \varepsilon'_R/\varepsilon'_T$ and $k_{2(R/T)} = tg\delta_R/tg\delta_T$ of oven-dry wood

Wood species	Relative volume of rays (%) (Borovikov and Ugolev 1989)	$k_{1(R/T)}$				$k_{2(R/T)}$			
		10^2–10^8 Hz		10^9–10^{11} Hz		10^2–10^8 Hz		10^9–10^{11} Hz	
		Limits	Average	Limits	Average	Limits	Average	Limits	Average
Spruce	5–7.2								
Pine	5.3–8.4								
Pseudothuga	7.3	1.07–1.02	1.05	1.03–1.01	1.02	1.09–1.03	1.06	1.05–1.02	1.04
Sequoia	7.8								
Larch and other conifers	8.8–10								
Birch	10.8–11.7								
Aspen	11	1.05–1.02	1.04	1.02–1.01	1.02	1.11–1.06	1.09	1.06–1.03	1.05
Elm	11–12								
Ash, chestnut	12								
Oak	22–28	1.11–1.05	1.08	1.05–1.03	1.04	1.25–1.12	1.19	1.14–1.05	1.10
Beech	27								

and

$$\varepsilon'_T = \frac{2\varepsilon'_\perp}{1 + k_{1(k/T)}} = \frac{2 \cdot 2.1}{1 + 1.04} = 2.06 \ .$$

The values for $tg\delta_R$ and $tg\delta_T$ are determined in a similar manner.

With a rise of frequency, the difference between the parameters of oven-dry wood in the radial and tangential directions diminishes.

It has been found experimentally that at a vector $\bar{E}$ orientation perpendicular to the fiber direction, the $\varepsilon'_\perp$ and $tg\delta_\perp$ values of oven-dry wood are influenced by tree species considerably less than by wood density. The density of the wood is the principal factor determining the dielectric properties of oven-dry wood, other conditions being the same. With increasing wood density, the ε' and $tg\delta$ values increase throughout the entire frequency range (Fig. 4.1). The maximum value of these dielectric parameters is at a wood density value equal to the

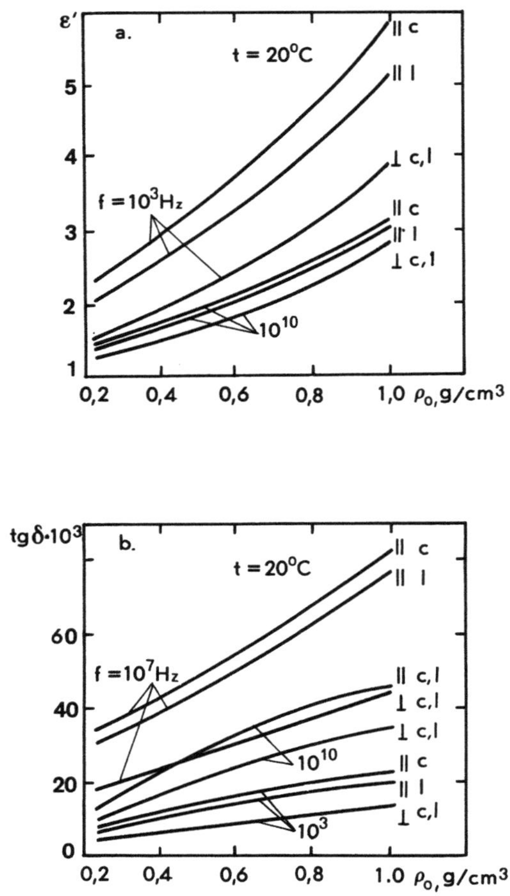

Fig. 4.1a, b. Dependence of the dielectric constant (**a**) and of loss tangent (**b**) of oven-dry wood on the wood density (ρ_0) at 20 °C: f frequency; c softwood; l hardwood; $\parallel$, $\perp$ vector $\bar{E}$ is parallel or perpendicular, respectively, to the longitudinal axis of anisotropy L

density of the cell wall substance. When vector E is oriented parallel to the fiber direction, the difference between $\varepsilon'_{\parallel}$ and $tg\delta_{\parallel}$ values of coniferous and angiosperm species of wood is considerable, but it diminishes with increasing frequency (Fig. 4.1). At frequencies exceeding 10^{10} Hz, this difference is near the error magnitude of the measurements. The influence of the orientation of vector $\bar{E}$ on the ε' and ε'' values of oven-dry softwoods and hardwoods in the LT plane is shown in Fig. 4.2

In addition to the gross structure of the wood, the anisotropy of the dielectric parameters of wood is equally affected by the microstructure of the cell walls. According to Norimoto (1976), the average angle of inclination of softwood microfibrils relative to the longitudinal axis of the cell in the radial wall of the earlywood cell is 23.5°–35.6°, and that of the late wood is 4.2°–14.2°. In the tangential wall of the cell, it is 12.5°–28.5° and 8.5°–19.7° respectively. The polar groups of cellulose and hemicelluloses have more freedom of movement along the fiber direction than in the transverse direction. This explains why oven-dry wood possesses larger values of the dielectric parameters when vector $\bar{E}$ is oriented in the longitudinal direction.

Figure 4.3 presents the frequency dependence of ε'' of oven-dry wood of *Magnolia obovata* along the three axes of anisotropy, and of cellulose and lignin. Norimoto and Yamada (1972) concluded that the dielectric properties of wood in the longitudinal direction are determined by the properties of cellulose as the maximum values of the loss factor of cellulose and that of wood occur at the same frequency. At the same time, lignin has a considerable influence on the properties of wood in the transverse direction as the frequency corresponding to its maximum values of dielectric loss (ε''_R and ε''_T) is shifted to the higher range in

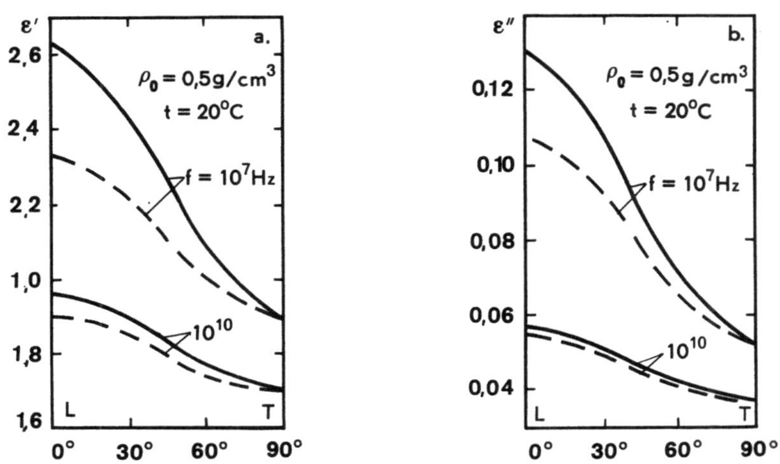

Fig. 4.2a, b. Curves showing changes of the dielectric constant ε' (**a**) and loss factor ε'' (**b**) of oven-dry wood as a function of the angle between vector $\bar{E}$ and direction L in the plane LT (*solid line* is for softwood, and *dotted line* is for hardwood)

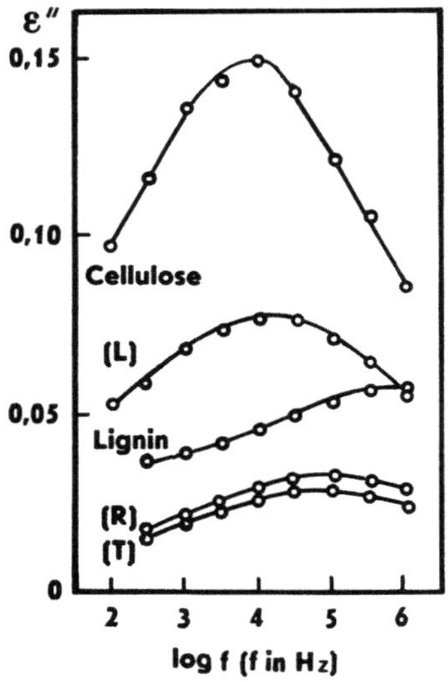

Fig. 4.3. Frequency dependence of the loss factor ε'' of cellulose, lignin, and *Magnolia obovata* wood in the longitudinal (L), radial (R), and tangential (T) directions at − 58 °C (Norimoto and Yamada 1972)

comparison with the frequency corresponding to the maximum value of the cellulose loss factor.

An increase in the proportion of latewood in the sample results in a corresponding increase of the dielectric constant (DC) value, which is connected with an increase of the average wood density (Table 4.2). An increase of the latewood proportion from 10 to 30% increases the DC value by 12–25%.

The empirical dependence of the loss factor of oven-dry wood on the density (varying from 0.1 to 0.8 g/cm^3) at a frequency 10^6 Hz and with a vector $\bar{E}$ orientation along the fiber axis was obtained by Norimoto (1976) in a study of 10 coniferous and angiosperm species:

At a temperature of 31 °C

$$\varepsilon''_{\parallel} = 0.284\,\rho_0 - 0.015\,. \tag{4.7}$$

At the temperature $t = -51$ °C

$$\varepsilon''_{\parallel} = 0.141\,\rho_0 - 0.0017\,. \tag{4.8}$$

Table 4.2. Relations between dielectric constant values and the proportion of latewood in coniferous species (Norimoto et al. 1978) ($W = 0\%, f = 10^6$ Hz, $t = 20$ °C)

Content of latewood	10%	20%	30%
Density (g/cm^3)	0.43	0.5	0.58
ε'_L	2.24	2.48	2.71
ε'_R	1.75	1.85	1.96
ε'_T	1.66	1.85	2.07

The correlation between the density and the loss factor values was very close. The correlation coefficient at 31 °C was equal to 0.967, and at − 51 °C it was 0.974.

Experimental data concerning the dielectric parameters of oven-dry wood obtained by various investigators for more than 60 coniferous and angiosperm species with densities ranging from 0.13 (balsa) to 1.53 g/cm³ (cell wall substance) were processed by computer. The correlation between $\varepsilon'_\perp$, $tg\delta_\perp$, and $\varepsilon''_\perp$ and the density was determined. These correlations, found with the vector of the electric field strength $\bar{E}$ oriented transverse to the fiber direction and at a temperature of 20 °C, were expressed by polynomials of the type $\varepsilon'_\perp = a_0 + a_1\rho_0 + a_2\rho_0^2$. Based on these correlations, tables with the dielectric parameter values of oven-dry wood and of cell wall substance for frequencies ranging from 10^2 to 10^{11} Hz were compiled. These tables are given in Appendix 1. The values of the static DC $\varepsilon'_{\perp s}$ and the limiting high-frequency DC $\varepsilon'_{\perp\infty}$ (the sign ∞ is conventionally accepted for a frequency equal to 10^{15} Hz) in these tables were derived by plotting Cole-Cole circle diagrams.

The data obtained by the majority of investigators differ from the $\varepsilon'_\perp$ values presented in Appendix 1 within ± 12%, and from $tg\delta_\perp$ values within ± 26%. The larger deviations of the $tg\delta_\perp$ values can be explained by the higher coefficient of variation (up to 27%) and larger measurement errors.

Figures 4.4, 4.5, and 4.6 show the graphical dependencies of the dielectric parameters of wood of various densities upon frequency. The oven-dry wood acts in an electromagnetic field as a typical polar dielectric. A dispersion of DC takes place. With increasing frequency, the DC monotonously diminishes, as a lesser part of the polar wood molecules have sufficient time to follow the external field changes. With the increase of wood density, the DC increases proportionally.

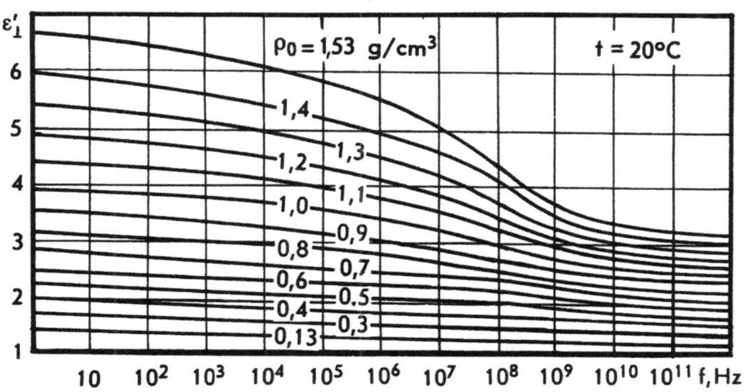

Fig. 4.4. Dielectric constant $\varepsilon'_\perp$ of oven-dry wood and of cell wall substance (ρ_0 = 1.53 g/cm³) as a function of frequency f with the vector of electric field strength oriented perpendicular to the fiber direction, at t = 20 °C

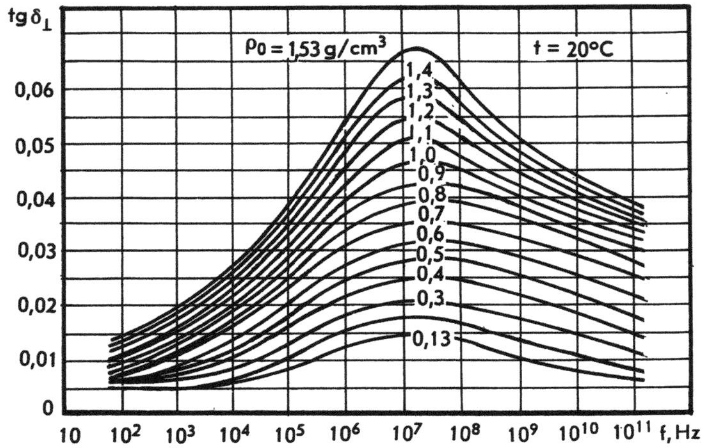

Fig. 4.5. Dependence of the loss tangent, $tg\delta_{\perp}$, of oven-dry wood and of the cell wall ($\rho_0 = 1.53$ g/cm^3) on frequency f with the vector of the electric field strength orientation perpendicular to the fiber direction and at $t = 20$ °C

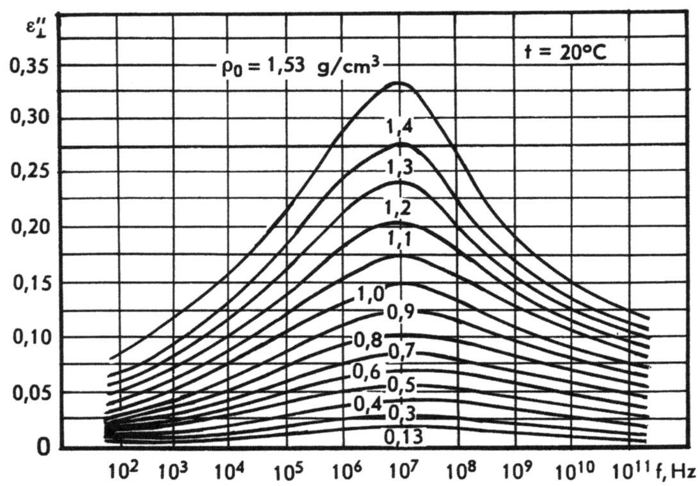

Fig. 4.6. Loss factor, $\varepsilon''_{\perp}$, of oven-dry wood and of cell wall substance ($\rho_0 = 1.53$ g/cm^3) as a function of frequency f with the vector of electric field strength oriented perpendicular to the fiber direction, at $t = 20$ °C

The loss tangent (Fig. 4.5) varies within 0.005–0.07, the maximum value being at frequencies around 10^7 Hz. The loss factor $\varepsilon''_{\perp}$ changes from 0.007 to 0.34, reaching the maximum value at 10^7 Hz (Fig. 4.6). The frequency corresponding to the maximum value of the loss factor of over-dry wood coincides with the frequency at which the maximum value of the loss factor of cellulose

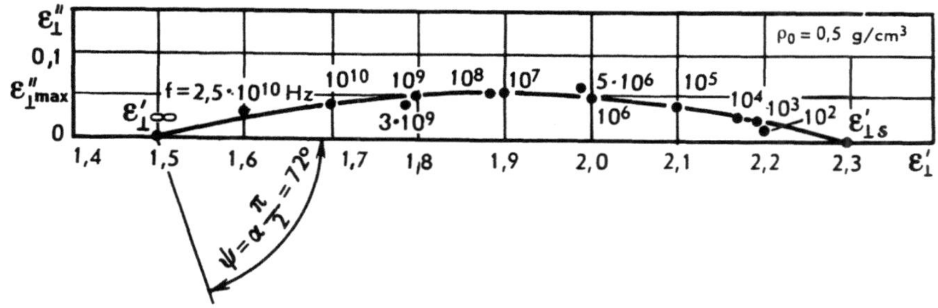

Fig. 4.7. Cole-Cole empirical diagram for oven-dry wood of density $\rho_0 = 0.5$ g/cm^3, at 20 °C: f frequency; $\varepsilon'_\perp$ dielectric constant; $\varepsilon'_{\perp s}$ static dielectric constant; $\varepsilon'_{\perp\infty}$ optic dielectric constant; $\varepsilon''_\perp$ loss factor

occurs. In oven-dry wood, all types of polarization mentioned in Chap. 1 occur except for electrolytic polarization.

The definition of the analytical frequency dependency of the DC and the loss factor of oven-dry wood with various density values can be carried out with the aid of Eq. (1.22) and Cole-Cole diagrams (Chap. 1). The molecules in wood possess different relaxation times, which is well illustrated by the Cole-Cole empirical diagram for oven-dry wood in Fig. 4.7. Distribution of the relaxation time τ follows certain statistical laws that are expressed with sufficient approximation by a normalized distribution function of the following type (Netushil et al. 1959)

$$G(\tau) = \sin \alpha\pi / 2\pi\tau \left[\operatorname{ch}(1-\alpha) \ln \frac{\tau}{\tau_m} - \cos \alpha\pi \right], \tag{4.9}$$

where

τ_m = the most probable relaxation time corresponding to frequency f_m at which the maximum loss value occurs
α = the parameter of relaxation time distribution ($0 < \alpha < 1$)
$\operatorname{ch}(1-\alpha)$ = hyperbolic cosine of $(1-\alpha)$.

The dependence of ε'' on ε' in the region of maximum relaxation for wood as a polar material with a relaxation time distribution according to Eq. (4.9) is expressed by Eq. (1.22)

$$\varepsilon' - i\varepsilon'' = \varepsilon'_\infty + (\varepsilon'_S - \varepsilon'_\infty)/[1 + (i\omega\tau_m)^{1-\alpha}], \tag{4.10}$$

which is the circle equation with the central point below the X-axis in the graphic presentation (Fig. 4.7). The angle ψ determines the parameter of relaxation time distribution α

$$\alpha = 2\psi/\pi. \tag{4.11}$$

The angle ψ in the Cole-Cole diagrams reported by various investigators for oven-dry wood varies considerably depending on the orientation of vector

$\bar{E}$ relative to the fiber direction, on temperature, wood density, and errors in measurements of ε' and ε'', and also on errors in plotting the diagrams. In the temperature range from 60 to $-70\,°C$, angle ψ is 52°–74°, and the parameter α varies from 0.58 to 0.82. On lowering the temperature, the value of parameter α increases.

The determination of angle ψ carried out with the use of well-known experimental data for woods of various densities and with vector $\bar{E}$ oriented perpendicular to the fiber direction has shown that its value varies within 68° to 74° with an average of 72°. For oven-dry wood and cell wall substance, parameter α is 0.8 at a temperature of 20 °C. The same value of α can be accepted for the temperature range from $+20\,°C$ to $-70\,°C$. The closer the value of α is to unity, the weaker is the dependence of ε'' on the frequency, and its maximum value is less clearly shown, which is illustrated in the case of wood in Fig. 4.7.

Having analyzed real and imaginary factors of Eq. (4.10), we can define the dependencies of ε' and ε'' on the frequency

$$\varepsilon' = \varepsilon'_\infty + (\varepsilon'_s - \varepsilon'_\infty)\frac{1 + (\omega\tau_m)^{1-\alpha}\cdot\cos\dfrac{\pi(1-\alpha)}{2}}{1 + (\omega\tau_m)^{2(1-\alpha)} + 2(\omega\tau_m)^{1-\alpha}\cdot\cos\dfrac{\pi(1-\alpha)}{2}} \tag{4.12}$$

$$\varepsilon'' = (\varepsilon'_s - \varepsilon'_\infty)\frac{(\omega\tau_m)^{1-\alpha}\cdot\sin\dfrac{\pi(1-\alpha)}{2}}{1 + (\omega\tau_m)^{2(1-\alpha)} + 2(\omega\tau_m)^{1-\alpha}\cdot\cos\dfrac{\pi(1-\alpha)}{2}}, \tag{4.13}$$

where

ω = angular frequency, $\omega = 2\pi f$.

When $\alpha = 0$, Eqs. (4.12) and (4.13) are reduced to the Debye equations.

The maximum value of the loss factor $\varepsilon''_\perp$ of oven-dry wood is at a frequency of 10^7 Hz, and the most probable relaxation time is derived from

$$\tau_m = 1/2\pi f_m = 1.59\times10^{-8}\,\mathrm{s}\,. \tag{4.14}$$

Substituting values $\tau = 1.59\times10^{-8}$ s and the parameter $\alpha = 0.8$ into Eqs. (4.12) and (4.13), we obtain frequency dependencies of $\varepsilon'_\perp = \varphi_1(\omega)$ and $\varepsilon''_\perp = \varphi_2(\omega)$ for oven-dry wood of any density at 20 °C

$$\varepsilon'_\perp = \varepsilon'_{\perp\infty} + (\varepsilon'_{\perp s} - \varepsilon'_{\perp\infty})\frac{1 + 2.62\times10^{-2}\,\omega^{0.2}}{1 + 5.25\times10^{-2}\omega^{0.2} + 7.6\times10^{-4}\omega^{0.4}} \tag{4.15}$$

$$\varepsilon''_\perp = (\varepsilon'_{\perp s} - \varepsilon'_{\perp\infty})\frac{0.85\times10^{-2}\,\omega^{0.2}}{1 + 5.25\times10^{-2}\omega^{0.2} + 7.6\times10^{-4}\omega^{0.4}}\,. \tag{4.16}$$

Values of $\varepsilon'_{\perp s}$ and $\varepsilon'_{\perp\infty}$ for wood of a certain density are given in Appendix 1. The values of $\varepsilon'_\perp$ calculated with the aid of Eq. (4.15) in the frequency range from 10^2 to 10^{11} Hz differ from average experimental data by $\pm 7\%$. Calculation of the $\varepsilon''_{\perp}$ value from Eq. (4.16) in the frequency range up to 10^5 Hz reveals errors within $+30$ to $+60\%$, in the range from 10^5 to 10^8 Hz the errors lie within $+10$ to $+20\%$, and at frequencies ranging from 10^8 to 10^{11} Hz they do not exceed $\pm 10\%$.

The relation of the dielectric properties of wood in the longitudinal and transverse directions is characterized by the coefficients k_1, k_2, and k_3 in Eq. (1.34). These coefficients depend to a great extent on the frequency and are different for coniferous and angiosperm species. The influence of the wood density on these coefficients is negligible. Thus, with an increase of density from 0.35 to 0.8 g/cm^3, $k_1 = \varepsilon'_{\parallel}/\varepsilon'_{\perp}$ at a frequency equal to 10^6 Hz increases by 6%–7% (Norimoto 1976). Analysis of the influence of the temperature on k_1 revealed that in the frequency range from 10^3 to 10^8 Hz an increase in temperature from 5 to 90 °C increases the coefficient by 10–15%. Under similar conditions, changes of $k_2 = tg\delta_{\parallel}/tg\delta_{\perp}$ for softwood are inconspicuous.

Based on well-known data and results obtained for the extent of influence of the orientation of the electric field strength vector, the empiric values of k_1, k_2, and k_3 were obtained. These values, calculated for the stated frequency range for oven-dry softwood and hardwood with a density of 0.2–1.0 g/cm^3 and at a temperature of 20–25 °C are given in Table 4.3.

The values of the dielectric parameters of oven-dry wood with vector $\bar{E}$ oriented parallel to the fiber direction can be derived from equations

$$\varepsilon'_{\parallel} = k_1\varepsilon'_{\perp}, \quad tg\delta_{\parallel} = k_2 tg\delta_{\perp}, \quad \varepsilon''_{\parallel} = k_3\varepsilon'_{\perp}\,, \tag{4.17}$$

where values of $\varepsilon'_{\perp}$, $tg\delta_{\perp}$, and $\varepsilon''_{\perp}$ are to be taken from Appendix 1 and k_1, k_2, and k_3, from Table 4.3. In order to determine the dielectric parameters of wood at an arbitrary orientation of vector $\bar{E}$ in relation to the fiber direction, Eqs. (1.32), (1.33) or (1.36), (1.37) can be used. For example, let us determine the DC value ε'_x of oven-dry birch wood with a density of 0.6 g/cm^3 at a frequency of 10^6 Hz and a temperature of 20 °C when vector $\bar{E}$ is oriented at the angle $\alpha_1 = 30°$ to the

Table 4.3. Values of coefficients k_1, k_2, and k_3 for oven-dry wood ($\rho_0 = 0.2$ –1.0 g/cm^3, t = 20–25 °C)

Wood	Coefficient	Frequency (Hz)										
		0	10^2	10^3	10^4	10^5	10^6	10^7	10^8	10^9	10^{10}	10^{11}
Còniferous	k_1	1.64	1.60	1.54	1.50	1.46	1.42	1.38	1.32	1.20	1.16	1.14
	k_2	–	1.6	1.7	1.75	1.8	1.85	1.8	1.7	1.5	1.3	1.25[a]
	k_3	–	2.6	2.6	2.6	2.6	2.6	2.5	2.2	1.8	1.5	1.4[a]
Angiosperm	k_1	1.42	1.38	1.36	1.32	1.28	1.26	1.22	1.16	1.14	1.12	1.11
	k_2	–	1.4	1.45	1.5	1.55	1.70	1.65	1.50	1.4	1.3	1.25[a]
	k_3	–	1.9	2.0	2.0	2.0	2.1	2.0	1.7	1.6	1.5	1.4[a]

[a]Obtained by extrapolation.

longitudinal axis of anisotropy, L, assuming that the DC values in the radial and tangential directions are equal to each other. The operations are as follows:

1. From Appendix 1 we find that $\varepsilon'_\perp = 2.2$
2. From Table 4.3 we determine that for hardwood $k_1 = 1.26$
3. Using Eq. (1.36) the value of ε'_x is derived as $\varepsilon'_x = \varepsilon'_\perp(k_1 \cos^2\alpha_1 + \sin^2\alpha_1) = 2.2(1.26 \times \cos^2 30° + \sin^2 30°) = 2.63$.

The value of $tg\delta_x$ is to be found similarly, bearing in mind the given orientation of $\bar{E}$.

4.2 Influence of Temperature on Dielectric Properties

The temperature considerably affects the dielectric properties of wood. For estimation of this influence, the equation suggested by Debye, which correlates the relaxation time τ with molecular constants, should be considered

$$\tau = 4\pi r^3 \eta / K_b T\,, \tag{4.18}$$

where

r = radius of the molecular sphere
η = viscosity, i.e., the influence of molecular bond forces
K_b = Boltzmann's constant
T = absolute temperature.

This equation is considered to be strictly valid only for dilute solutions, but it makes it possible to provide a qualitative estimation of the temperature influence on the dielectric parameters of wood. An increase in temperature causes a reduction of τ. The decrease of relaxation time induces a shift of the maximum value of the wood loss tangent to the region of higher frequencies (Fig. 4.8). The

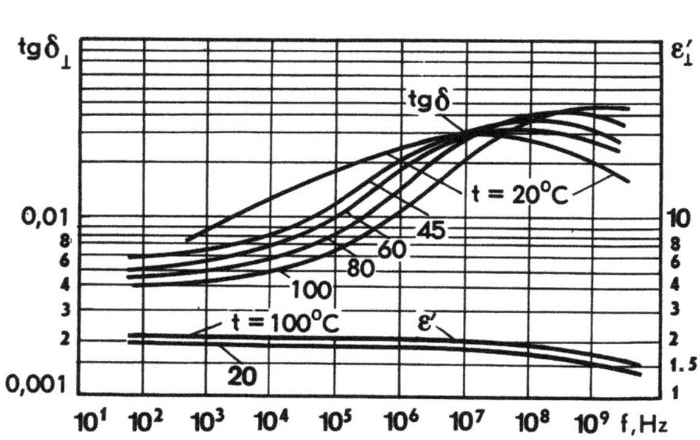

Fig. 4.8. Dependence of loss tangent $tg\delta_\perp$ and dielectric constant $\varepsilon'_\perp$ of oven-dry spruce wood on temperature and frequency (Trapp and Pungs 1956b)

Table 4.4. Most probable relaxation time τ_m of oven-dry wood and frequency f_m corresponding to the maximum value of the loss factor (Trapp and Pungs 1956b; Tsutsumi 1967)

	Temperature							
	−70 °C	−40 °C	−20 °C	20 °C	40 °C	60 °C	80 °C	100 °C
τ_m (s)	1.1×10^{-4}	1.6×10^{-6}	0.8×10^{-6}	1.6×10^{-8}	3.2×10^{-9}	1.3×10^{-9}	4×10^{-10}	1.6×10^{-10}
f_m (Hz)	1.4×10^{3}	1×10^{5}	2×10^{5}	1×10^{7}	5×10^{7}	1.2×10^{8}	4×10^{8}	1×10^{9}

higher temperature also results in a decrease of the viscosity, which in its turn contributes to the shift of not only the maximum point of the loss tangent but also the whole curve $tg\delta = \varphi(f)$ to higher frequencies.

The maximum value of the loss factor of oven-dry wood with decreasing temperature also shifts to a lower frequency (Hoffmann and Poliszko 1985) as the loss tangent, i.e., the relaxation time of the molecules, increases. Table 4.4 shows the most probable values of relaxation time τ_m and of frequencies f_m that correspond to the maximum values of the absorption of dry wood in the temperature range from −70 to 100 °C according to data by Trapp and Pungs (1956b) and Tsutsumi (1967).

The value of DC of oven-dry wood increases continuously with increasing temperature (Fig. 4.9). The values of the temperature coefficient of oven-dry wood DC, $k^t_{\varepsilon'_\perp}$, calculated on the basis of the data reported by Brown et al. (1947), Trapp and Pungs (1956b), Tinga (1969), Nanassay (1970), James (1975), Norimoto (1976), Benkova (1981), and Torgovnikov (1982) are as follows: in the temperature range from −40 to +100 °C and at frequencies up to 10^7 Hz the coefficient values are 1.0×10^{-3} to $5.5\times10^{-3}\,K^{-1}$, and at frequencies of 10^7–10^{10} Hz they range from 0.7×10^{-3} to $2.2\times10^{-3}\,K^{-1}$. The data offered by Nanassay (1970) are exceptional; he has reported values of $k^t_{\varepsilon'_\perp}$ at low frequencies from 8×10^{-3} to $12\times10^{-3}\,K^{-1}$.

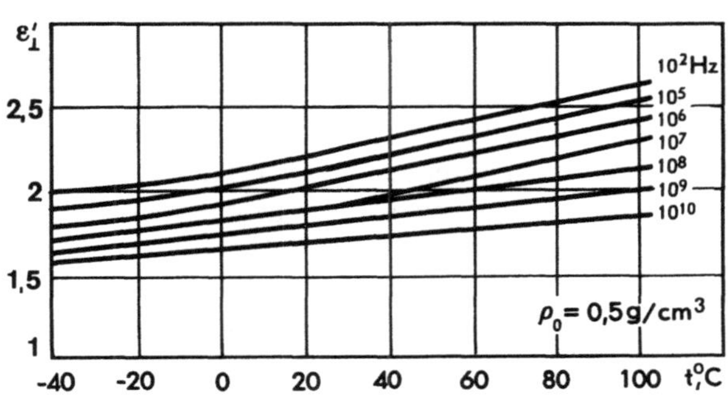

Fig. 4.9. Dependence of the dielectric constant of oven-dry wood on the temperature (wood density 0.5 g/cm^3)

Table 4.5. Temperature coefficient of the dielectric constant of oven-dry wood, $k^t_{\varepsilon'}$ $\times 10^3 K^{-1}$

Temperature range (°C)	Orientation of $\bar{E}$ relative to fiber direction	Frequency (Hz)		
		10^2–5×10^6	10^7–2.4×10^9	3×10^9–10^{11}
− 40 to + 20	⊥	1.5	1	1
	∥	2.3	1.5	1.5
21 to 100	⊥	2.5	1.5	1
	∥	3.8	2.2	1.5

My own research, carried out at frequencies from 1.0 to 3 GHz, gave a value of $k^t_{\varepsilon'_\perp} = 1.5\times10^{-3}\,K^{-1}$, and at frequencies from 4.0 to 7.5 GHz $k^t_{\varepsilon'_\perp}$ was $1.0\times10^{-3}\,K^{-1}$. When vector $\bar{E}$ is oriented parallel to the fiber direction, the temperature coefficient of the DC is 1.5 times higher than with a transverse orientation. The average values of the DC temperature coefficient are given in Table 4.5. The value of the temperature coefficient in the temperature range of 21–100 °C was higher than at temperatures varying from − 40 to + 20 °C.

The ε' value of wood at a specified temperature equal to T can be calculated from

$$\varepsilon' = \varepsilon'_{20}[1 + k^t_{\varepsilon'}(T - 293)] , \tag{4.19}$$

where

ε'_{20} = dielectric constant at $t = 20\,°C$.

The values of $\varepsilon'_{20\perp}$ can be taken from Appendix 1, the values of $\varepsilon'_{20\parallel}$ can be calculated from Eq. (4.17), and the values of $k^t_{\varepsilon'}$ are given in Table 4.5.

The loss tangent of oven-dry wood has a complicated dependence on the temperature (Fig. 4.10). With an increase in temperature from − 40 to + 100 °C at frequencies up to 10^5 Hz, the values of $tg\delta$ diminish. At a frequency of 10^6 Hz, $tg\delta$ has its maximum value in the temperature range from − 20 up to 0 °C and at a frequency of 10^7 Hz the maximum shifts to $t = 20\,°C$. With an increase of frequency to 10^8 Hz it moves further to the temperature range 80–100 °C. At frequencies up to 10^5 Hz, the temperature coefficient of the loss tangent $k^t_{tg\delta}$ is negative at temperatures above − 40 °C. At a frequency of 10^6 Hz it is negative at temperatures above 0 °C, and at a frequency of 10^7 Hz it is negative at temperatures exceeding 20 °C. At frequencies of 10^8–10^{10} Hz, the coefficient $k^t_{tg\delta}$ is positive in the whole range of temperatures from − 40 to + 100 °C.

The values of $k^t_{tg\delta}$ at positive temperatures up to 100 °C in the frequency range of 10^3–10^8 Hz vary from -11×10^{-3} to $+16\times10^{-3}\,K^{-1}$. At frequencies exceeding 10^8 Hz, the values of $k^t_{tg\delta}$ obtained by me were within 2.2×10^{-3} to $11\times10^{-3}\,K^{-1}$. In the negative temperature region, $k^t_{tg\delta}$ varies from -40×10^{-3} to $+8\times10^{-3}\,K^{-1}$. The influence of the orientation of vector

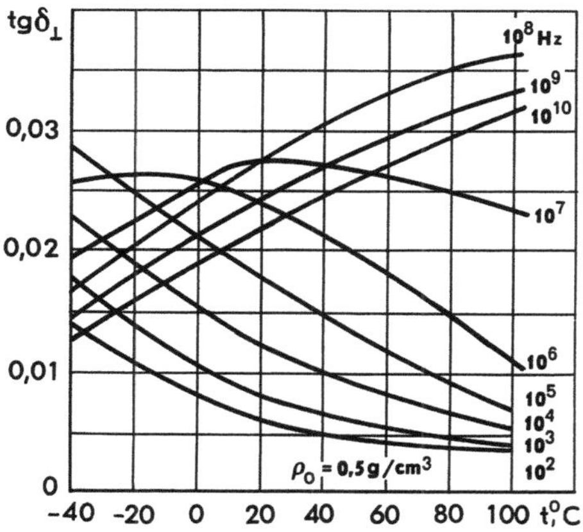

Fig. 4.10. Dependence of loss tangent of oven-dry wood on the temperature (wood density 0.5 g/cm³)

$\bar{E}$ relative to that of the fibers on $k^t_{tg\delta}$ has not been determined. The average values of the temperature coefficient of the loss tangent are given in Table 4.6.

The value of $tg\delta$ of wood at a specified temperature T can be found from

$$tg\delta = tg\delta_{20}[1 + k^t_{tg\delta}(T - 293)] , \quad (4.20)$$

where

$tg\delta_{20}$ = loss tangent value at 20 °C.

The values of $tg\delta_{20\perp}$ can be taken from Appendix 1, the values of $tg\delta_{20\parallel}$ can be calculated from Eq. (4.17), and the values of $k^t_{tg\delta}$ are found in Table 4.6.

The influence of temperature on the DC value and on the loss factor of oven-dry beech wood in the temperature range from − 196 to 177 °C is shown in Fig. 4.11 (Handa et al. 1982; Hoffmann and Poliszko 1985). The general character of the results obtained by various scientific workers is identical although the absolute values differ considerably. The maximum value of the loss factor at a frequency of 10^4 Hz and with vector $\bar{E}$ oriented parallel to the fiber direction occurs at temperatures from − 40 to − 50 °C. The process of lowering the frequency shifts the maximum value of $\varepsilon''_{\parallel}$ to the region of lower temperatures, and at a frequency of 30 Hz this maximum value is observed at the temperature of about −100 °C.

Table 4.6. Temperature coefficient values of the loss tangent of oven-dry wood $k^t_{tg\delta} \times 10^3$ K^{-1}

Temperature range (° C)	Frequency (Hz)								
	10^2	10^3	10^4	10^5	10^6	10^7	10^8	10^9	10^{10}
− 40 to + 20	− 20	− 20	− 15	− 10	− 1	+ 5	+ 7	+ 7	−
21 to 100	− 5	− 6	− 6	− 8	− 7	− 2	+ 4	+ 4	+ 6

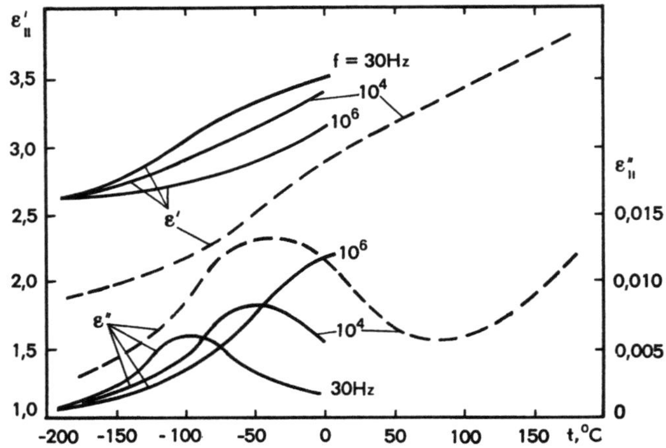

Fig. 4.11. Dielectric constant $\varepsilon'_{\parallel}$ and loss factor $\varepsilon''_{\parallel}$ of oven-dry beech wood as a function of temperature, with the vector of electric field strength orientation parallel to the fiber direction: f frequency; t temperature (*solid lines* are plotted according to data of Handa et al. 1982, *dotted lines* according to data of Hoffmann and Poliszko 1985)

The influence of the field strength on the dielectric properties of wood has not been sufficiently studied yet, a fact that is probably explained by the very small magnitude of this influence. According to the studies of Nanassy (1972), an increase of the electric field strength to 434 V/cm within a frequency range of 10^2–10^5 Hz does not affect the values of ε' and $tg\delta$ of the oven-dry wood.

Chapter 5

Dielectric Properties of Moist Wood

The absolute value of the moisture content of wood can reach 250% and even higher. Free moisture has values of dielectric properties many times higher than those of wood. Interacting with wood in the process of adsorption and capillary condensation, the moisture changes its own dielectric properties and causes changes in the dielectric properties of the wood. Free moisture filling the cell cavities also changes the character of interactions between the wood and the electromagnetic field. These changes result in changes of the dielectric properties of the wood, particularly at frequencies at which water possesses anomalously high dielectric parameters. Water is the ingredient that changes the dielectric properties of wood the most and in many cases determines them.

The physical factors that determine the character of the influence of moisture on the dielectric properties of wood have not yet been sufficiently studied theoretically. The results of these studies can give only a qualitative estimation of the process. This is the reason why the data adduced below on the influence of moisture on the dielectric properties of wood are mainly based on the results of experimental research.

5.1 The Dielectric Constant of Moist Wood

At low frequencies in the range from 10 to 50 Hz, at a moisture content of up to 5%, and at a temperature of 20–25 °C, when wood contains only monolayer moisture, the dielectric constant (DC) rises slowly and reaches values from 3–4 to 8–12, at $W = 4\%–6\%$. These figures are valid with an orientation of vector $\bar{E}$ perpendicular to the fiber direction. When vector $\bar{E}$ is oriented parallel to the fibers, the values of DC are approximately twice as high (Trapp and Pungs 1956b, Tsutsumi 1967, Beldi et al. 1968, Maeda et al. 1977, Maeda and Fukada 1987). An increase of moisture content from 5 to 10% increases the DC 3–10 times. A further increase of the moisture content increases the DC by hundreds and thousands of times.

In the frequency range up to 10^3 Hz and at a moisture content above 10%, the DC values given by various authors differ among themselves tenfold and more, indicating the existence of great variations and difficulties in measuring the DC values in this frequency range. The results obtained by Mikhailovskaja (1972) showed that the DC value of Siberian larch wood at a frequency of 50 Hz, $t = 20$ °C, and $\rho_0 = 0.62$ g/cm^3 with a rise of moisture from 10 to 70% increased 200–400 times, the orientation of vector $\bar{E}$ being perpendicular to the fiber direction. If vector $\bar{E}$ is parallel to the fiber, this increase reaches 2400 times. At

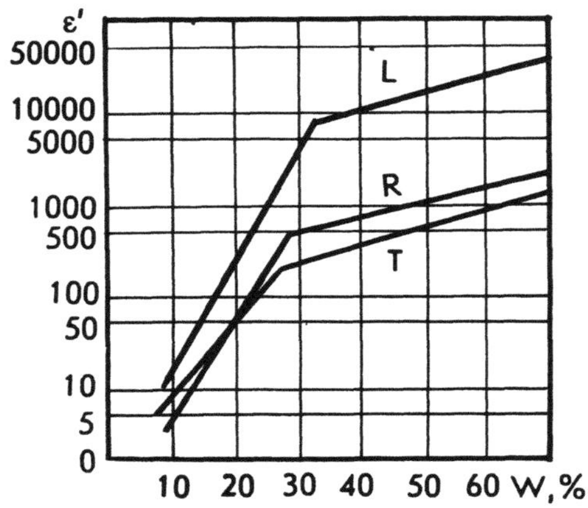

Fig. 5.1. Dielectric constant of Siberian larch wood at a frequency of 50 Hz ($\rho_0 = 0.62$ g/cm^3, $t = 20\,°C$), according to data of Mikhailovskaja (1972), and with vector $\bar{E}$ oriented in longitudinal (L), radial (R), and tangential (T) directions

a moisture content which exceeds 15%, $\varepsilon'_R > \varepsilon'_T$, and ε'_L always exceeds the value of ε'_R (Fig. 5.1). The correlation coefficient of ε' and moisture content of larch wood at a moisture content up to 30% is within 0.93–0.98, and at a moisture content from 30 to 70%, it is within 0.69–0.80.

The values of ε' of Siberian larch with any moisture content at $t = 20\,°C$ and at a frequency of 50 Hz can be determined from the following empirical equations:

Moisture content range (%)	
8–29	$\varepsilon'_R = 0.644 \times (1.26)^W$
30–70	$\varepsilon'_R = 186 \times (1.04)^W$
8–26	$\varepsilon'_T = 0.94 \times (1.23)^W$
27–70	$\varepsilon'_T = 69.63 \times (1.046)^W$
8–33	$\varepsilon'_L = 1.16 \times (1.31)^W$
34–70	$\varepsilon'_L = 2\,663 \times (1.04)^W$

where

W = wood moisture content (%).

The DC of softwood (James 1975) on changing the moisture content from 10 to 120–130% at $t = 25\,°C$ and $f = 20$ Hz increases when the orientation of vector $\bar{E}$ is perpendicular to the fiber direction from 20–24 to 26 000–30 000, and with an orientation of $\bar{E}$ parallel to the fibers from 60–95 to 300 000–430 000, i.e., by thousands of times. At a frequency of 10^2 Hz, $\varepsilon'_\perp$ changes from 12–14 to 3500–3700, and $\varepsilon'_\parallel$ changes from 30–40 to 40 000–50 000. At a frequency of 10^3 Hz, these changes are as follows: $\varepsilon'_\perp$ from 6–8 up to 300–320, and $\varepsilon'_\parallel$ from 18–20 to 2500–2700. At the same time, ε'_R exceeds ε'_T by 3–20%.

The DC of white oak wood is higher than that of softwood (James 1975), and at a moisture content equal to 10–80% and a frequency of 20 Hz, ε'_R changes from 40 to 31 000, ε'_T from 30 to 8000, and ε'_L from 200 to 700 000. At a frequency of 10^2 Hz, the changes are the following: ε'_R 15 to 5000, ε'_T 20 to 2000, ε'_L 100 to 100 000; at a frequency of 10^3 Hz, these changes are: ε'_R 12 to 1300, ε'_T 9 to 700, and ε'_L 40 to 9500.

The high value of the DC at low frequencies is due to the fact that moist wood has a high ionic conduction. The higher DC value of oak wood in longitudinal and radial directions in comparison with that of softwood can be explained by the high proportion of vessels and rays in oak wood.

At frequencies up to 10^3 Hz and at a moisture content exceeding 10%, the values of DC depend largely on the tree species and the density of the wood. These values differ considerably from one another, and their dependence on the moisture content has the character of an exponential function. Figure 5.2 shows

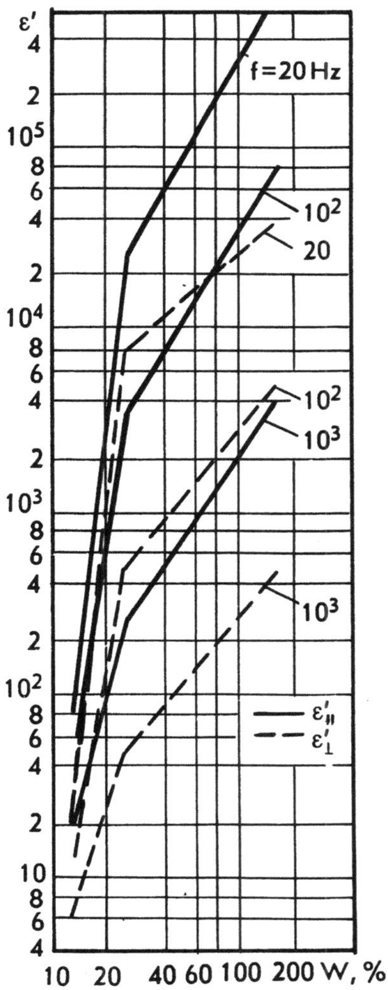

Fig. 5.2. Dependence of the softwood dielectric constant on moisture content in the frequency range from 20 to 1000 Hz ($\rho_0 = 0.4$–0.5 g/cm^3, $t = 20$–$25\,°C$)

Table 5.1. Formulae for calculation of the dielectric constant of moist wood at frequencies of 20–1000 Hz ($t = 20$–$25\,°C$)

f (Hz)	W (%)	ε'_R	ε'_T	ε'_L
Coniferous wood ($\rho_0 = 0.4$–$0.55\ g/cm^3$)				
20	25–120	$74.76\,W^{1.23}$		$74.32\,W^{1.81}$
10^2	12–24	$5.74 \times 10^{-5}\,W^{4.99}$		$7.93 \times 10^{-6}\,W^{6.21}$
	25–120	$10.52\,W^{1.24}$		$15.34\,W^{1.67}$
10^3	12–24	$1.94 \times 10^{-2}\,W^{2.41}$		$4.18 \times 10^{-3}\,W^{3.42}$
	25–120	$0.77\,W^{1.27}$		$1.63\,W^{1.55}$
Oak ($\rho_0 = 0.65$–$0.75\ g/cm^3$)				
20	25–100	$77.12\,W^{1.37}$	$190.6\,W^{0.86}$	
10^2	12–24	$3.63 \times 10^{-5}\,W^{5.34}$	$6.41 \times 10^{-5}\,W^{4.98}$	
	25–100	$14\,W^{1.35}$	$20.7\,W^{1.04}$	
10^3	12–24	$1.15 \times 10^{-3}\,W^{3.72}$	$2.25 \times 10^{-3}\,W^{3.35}$	$1.73 \times 10^{-3}\,W^{4.01}$
	25–100	$0.81\,W^{1.69}$	$0.61\,W^{1.61}$	$0.49\,W^{2.25}$

W = wood moisture content (%)

that, with an increase in moisture content to 25–30%, the rate of increase in the DC is higher than at a higher moisture content.

The equations that make it possible to determine the DC of moist coniferous wood and oak wood at low frequencies are given in Table 5.1 in which the values of W are expressed in percentage. These equations were produced on the basis of data obtained by Trapp and Pungs (1956b), Beldi et al. (1968), Nanassy (1972), and James (1975). Appendix 2 contains data on the dielectric properties of wood at frequencies from 20 to 1000 Hz and in the temperature range from -20 up to $+90\,°C$, as reported by James (1975).

At a moisture content below 10–15% and in the frequency range from 20 to 10^3 Hz, as well as at a moisture content up to 30% and at frequencies above 10^3 Hz, the influence of the tree species on the value of $\varepsilon'_\perp$ is negligible, provided the wood density is the same. The decisive factors influencing $\varepsilon'_\perp$ are the moisture content and the density of the wood (Skaar 1948; Hearmon and Burcham 1954, Peterson 1960; Norimoto 1976; Peyskens et al. 1984). At a moisture content up to 30%, the correlation coefficient between $\varepsilon'_\perp$ and the oven-dry wood density ρ_0 is within 0.85–0.99, but in most cases it is above 0.95. An analysis of the experimental data showed that at frequencies above 10^3 Hz and at a moisture content below 30%, the value of $\varepsilon'_\perp$ has a linear dependence on the wood density (Fig. 5.3), the family of straight lines of $\varepsilon'_\perp = \varphi(\rho_0)$ passing the point $\rho_0 = 0$, $\varepsilon' = 1$ (DC of a vacuum).

At a certain level of moisture content in the range from zero to 30% and at a temperature of 20–25 °C, the dependence of $\varepsilon'_\perp$ on the wood density (within the limits of $\rho_0 = 0.3$–$0.8\ g/cm^3$), is expressed by empirical equations of the following type

$$\varepsilon'_\perp = a\rho_0 + 1\,, \tag{5.1}$$

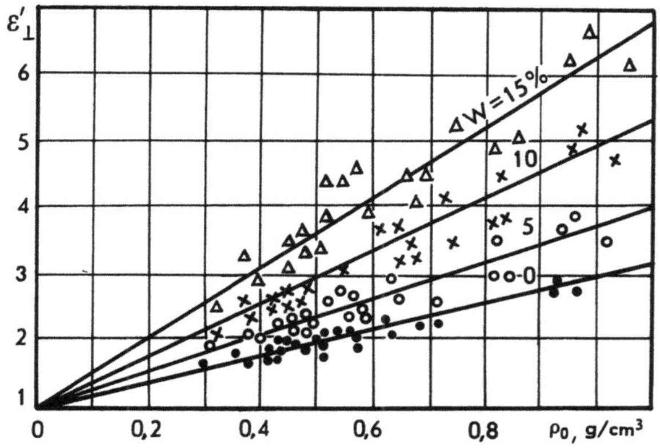

Fig. 5.3. Dielectric constant of moist wood as a function of wood density at a frequency $f = 2 \times 10^6$ Hz and at $t = 20\,°C$; moisture content W (%) (Skaar 1948)

where

ρ_0 = density of oven-dry wood (g/cm³)

a = coefficient, the value of which depends on frequency and moisture content, as is shown in Table 5.2.

The dependence of the values of $\varepsilon'_\perp$ of moist wood upon wood density, moisture content, and frequency is shown in Appendix 3.

Deviations of experimental $\varepsilon'_\perp$ values obtained with wood of a moisture content of 5–30% from values calculated from Eq. (5.1) and with coefficients

Table 5.2. Coefficient a value in Eq. (5.1) ($t = 20$–$25\,°C$, $\rho_0 = 0.3$–0.8 g/cm³)

f (Hz)	Moisture content (%)					
	5	10	15	20	25	30
20	4.6	35.8				
10^2	4.2	19.6				
10^3	3.9	10.2	28.2			
10^4	3.8	6.4	13.4	22.3	31.6	42.4
10^5	3.7	5.9	8.9	14.3	20.8	31.5
10^6	3.0	4.2	5.8	8.1	11.2	15.0
10^7	2.9	4.0	5.6	7.4	9.3	11.4
10^8	2.6	3.7	5.2	6.2	7.3	8.2
10^9	2.2	2.9	3.8	4.8	5.9	7.3
2.4×10^9	1.9	2.5	3.2	3.8	4.6	5.5
10^{10} [a]	1.5	1.9	2.4	2.8	3.6	4.7
10^{11} [a]	1.3	1.6	1.8	2.2	2.6	3.2

[a] Result of extrapolation

a taken from Table 5.2 are as follows:

- at frequencies up to 10^5 Hz they are within $\pm$ 20%.
- at frequencies from 10^6 to 10^8 Hz they are within $\pm$ 18%.
- at frequencies between 10^9 and 10^{10} Hz they are within $\pm$ 15%.

However in some cases, at a moisture content of 20–30%, these deviations can reach values of 24–45% (Beldi et al. 1968, James 1975, Torgovnikov 1982).

For the determination of $\varepsilon'_\perp$ at frequencies between those listed in Table 5.2, a diagram showing the dependence of coefficient a on frequency and moisture content (Fig. 5.4) can be used. Having determined the value a from the diagram under the specified conditions and given wood density, the value of the DC can be found by using Eq. (5.1).

In dry wood, $\varepsilon'_R > \varepsilon'_T$, and this inequality is valid for moist wood as well. In those cases when it is not sufficient to know the DC value in the transverse direction and the difference betwen ε'_R and ε'_T should be taken into consideration, the values of ε'_R and ε'_T can be determined from Eq. (4.5) with the aid of coefficients $k_{1(R/T)}$ from Table 5.3.

At a moisture content below 30%, oak wood and other woods with more than 18 volume per cent rays show $k_{1(R/T)}$ values within 1.1–2.1 at frequencies up to 10^5 Hz. These figures exceed those of wood species with a lower volume of rays (1–1.3). With an increase of frequency, the difference between ε'_R and ε'_T diminishes, and the frequencies of 10^9 Hz and higher it is not detectable.

It should be noted that, on analysis of the theoretical data reported by Norimoto et al. (1978), an increase of the relative volume of latewood above 20–25% in a softwood results in a situation where the ε'_T value of moist wood can be higher than ε'_R. It is possible that the samples of Douglas-fir wood that

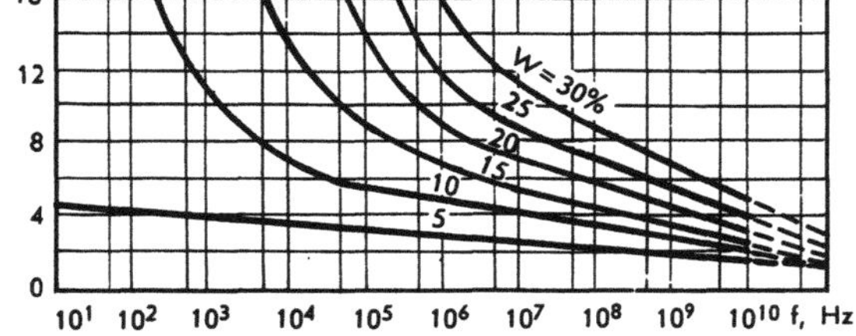

Fig. 5.4. Dependence of coefficient a on frequency f and moisture content W, at t = 20–25 °C

Table 5.3. Coefficient $k_{1(R/T)} = \varepsilon'_R/\varepsilon'_T$ for moist wood at $t = 20–25\,°C$ and density 0.3–0.8 g/cm^3

f (Hz)	Species	Moisture content (%)						
		5	10	15	20	25	30	40 and more
$20–10^3$	Oak, beech, and species with a ray volume exceeding 18%	1.1–1.15	1.3–1.4	1.4–1.6	1.5–1.9	1.6–2.1	1.7–2.1	1.8–2.6
$10^4–10^5$	Similar to above	1.1	1.1	1.1–1.2	1.3–1.5	1.5–1.9	1.7–2.1	–
$10^4–10^5$	Pine, spruce, Douglas-fir, birch, aspen, and species with a ray volume of 5%–12%	1–1.1	1–1.1	1.1	1.1	1.1–1.2	1.1–1.3	1–1.3
$10^6–10^8$	All species	1–1.1	1–1.1	1.1	1.1–1.15	1.1–1.2	1.15–1.25	1
$10^9–10^{10}$	All species	1	1	1	1	1	1	1

were used by James (1975) had a very high percentage of latewood. This may be an explanation of the fact that according to his data in many of his experiments the value of ε'_T exceeded the value of ε'_R. At frequencies of 10^6 Hz and higher, no significant difference in the values of coefficient $k_{1(R/T)}$ of softwood and hardwood could be ascertained. The average values for the dependence of $k_{1(R/T)}$ on the moisture content and frequency at a temperature of 20–25 °C are given in Table 5.3. From an analysis of the data by James (1975), the conclusion can be drawn that at positive temperatures up to 90 °C and moisture contents of 5–30%, the influence of the temperature on the value of coefficient $k_{1(R/T)}$ for softwood is negligible. However, for oak wood under the same conditions but at $W = 5$–20% and at $f = 10^4$ Hz and higher, the value of $k_{1(R/T)}$ increases by 6–20% with increasing temperature.

For the determination of $\varepsilon'_\perp$ of wood with a moisture content above 30%, Eq. (2.4) can be used

$$\varepsilon'_\perp = \varepsilon'_{p\perp} + V_4(\varepsilon'_w - 1)\,, \tag{5.2}$$

where

$\varepsilon'_{p\perp}$ = DC of wood with a moisture content of 30%, with vector $\bar{E}$ oriented perpendicular to the fiber direction; the value of DC can be determined from Eq. (5.1) or taken from Appendix 3

V_4 = relative volume of free moisture in the wood, taken from Table 2.1

ε'_w = DC of free moisture.

A value of $\varepsilon'_\perp$ calculated with the use of Eq. (5.2) satisfactorily agrees with the experimental data at a moisture content of 40%–50% in the frequency range of 10^4–10^9 Hz as well as at a moisture content of 40–120% and at frequencies of 10^6–10^7 Hz. At frequencies of 10^4–10^5 Hz and a moisture content $W > 50\%$, the calculated values of $\varepsilon'_\perp$ are considerably lower than the experimental ones and do not exceed the value of the DC of water. At the same time, experimental data give a value equal to 80–120. These results can be explained by the action of the ionic conduction of water in the wood. In the frequency range from 10^8 to 10^{10} Hz, the calculated values of DC exceed the experimental ones by 1.5–2.1 times. Therefore, for wood with a moisture content of 60–120%, and at a temperature of 20–25° C, an experimental correction factor "c" (see table 5.4) is introduced into Eq. (5.2)

$$\varepsilon'_\perp = [\varepsilon'_{p\perp} + V_4(\varepsilon'_w - 1)]/c\,. \tag{5.3}$$

A comparison of calculated [from Eq. (5.3)] and experimental data for the DC

Table 5.4. Values for correction factor c in Eq. (5.3) used for calculation of $\varepsilon'_\perp$ values of wood with a moisture content of 60–120%

Frequency (Hz)	10^4	10^5	10^6	10^7	10^8	10^9	10^{10}
c	0.5	0.7	1	1	1.5	2.1	2.1

shows a satisfactory agreement (Fig. 5.5). The average values of $\varepsilon'_\perp$ for wood with a density of 0.3–0.8 g/cm^3 and a moisture content of 60% and 100% and at frequencies from 10^4 to 10^{10} Hz are given in Appendix 3.

The dispersion (frequency response) of $\varepsilon'_\perp$ of moist wood (with $\rho_0 = 0.5$ g/cm^3) is given in Fig. 5.6. An abrupt decrease of DC in the range 10^{10}–10^{11} Hz is explained by a decrease of the value of the DC of free water

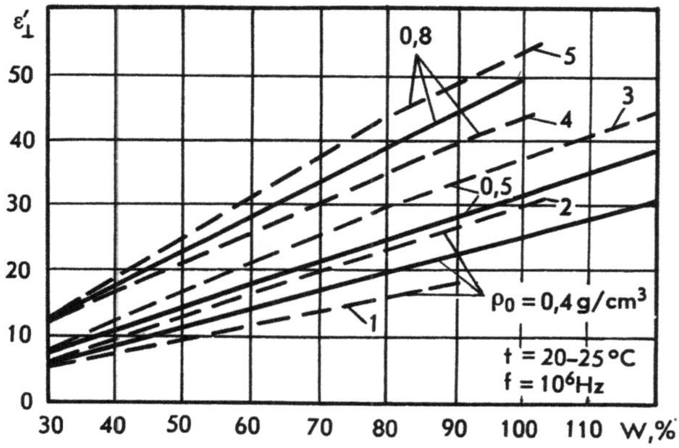

Fig. 5.5. Calculated (*solid lines*) and experimental (*dotted lines*) curves of dependence of the dielectric constant of wood on moisture content (above 30%) with vector $\overline{E}$ oriented transverse to the fibers, according to data of: *1* Goriajev (1975); *2* Trapp and Pungs (1956b); *3, 4* James (1975); *5* Uyemura (1960)

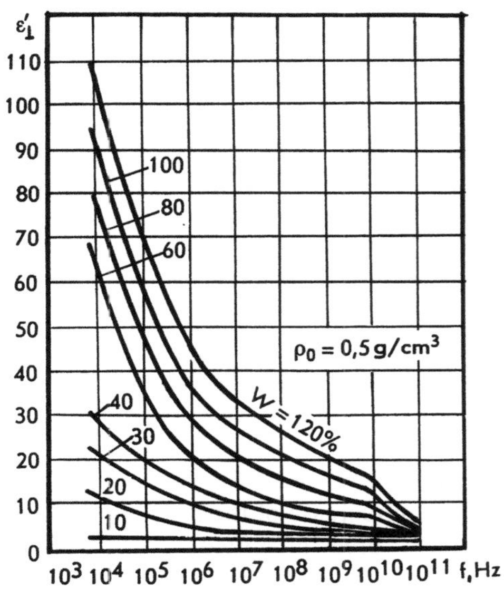

Fig. 5.6. Dielectric constant $\varepsilon'_\perp$ of moist wood in the frequency range from 10^4 to 10^{11} Hz at 20 °C: *f* frequency; *W* moisture content

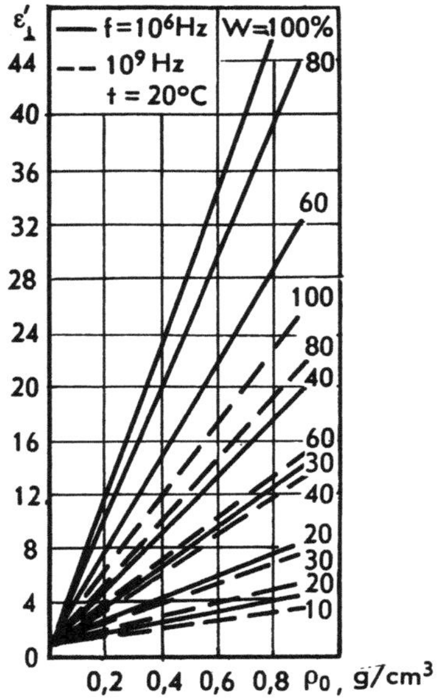

Fig. 5.7. Dependence of the dielectric constant $\varepsilon'_\perp$ on wood density and moisture content at frequencies of 10^6 and 10^9 Hz and temperature $t = 20\,°C$: ρ_0 density value of oven-dry wood; W moisture content

from 60 to 7.5. The dependence of the DC of moist wood upon the density of oven-dry wood is given in Fig. 5.7.

The DC $\varepsilon'_{||}$ of wood with the orientation of vector $\bar{E}$ parallel to the fibers is always higher than that with the perpendicular orientation ($\varepsilon'_\perp$) of vector $\bar{E}$. With increasing frequency, the relation $k_1 = \varepsilon'_{||}/\varepsilon'_\perp$ decreases. The variation of the k_1 values is very extensive, and no definite influence of tree species and density on the k_1 value has been established. An analysis of experimental data showed that at indoor temperatures, with the moisture content changing from 5 to 120%, and at frequencies from 20 to 10^2 Hz, coefficient k_1 changes from 1.8 to 28; at frequencies from 10^3 to 10^4 Hz, these limits are 1.4–9.3, and at frequencies from 10^5 to 10^{10} Hz, they are 1.1–2.9. In this case, at $t = 20$–$25\,°C$ and at frequencies of 20–10^4 Hz, the value of k_1 increases with increasing moisture content. The maximum value of k_1 appears at frequencies of 10^5–10^8 Hz at moisture contents from 25 to 40%. At frequencies of 10^9 Hz and higher, the maximum value of k_1 is reached at a moisture content of 25–30%. Upon further increase of the moisture content, the value of k_1 remains practically unchanged. The average values of coefficient k_1 as a function of moisture content and frequency are given in Table 5.5. In order to determine the $\varepsilon'_{||}$ of wood under the specified conditions, it is necessary to find the value of $\varepsilon'_\perp$ from Appendix 3 or to calculate it from Eqs. (5.1), (5.2), and (5.3) and to multiply by the coefficient k_1 taken from Table 5.5.

The relative volume of earlywood and latewood affects the value of the DC significantly, a fact that is largely caused by an increase of the average value of the density of the wood. Table 5.6 shows DC values of softwood with relative volume of latewood equal to 10 and 30%, at a frequency of 10^6 Hz.

Table 5.5. Average values of the coefficient $k_1 = \varepsilon'_{\parallel}/\varepsilon'_{\perp}$ (t = 20–25 °C, ρ_0 = 0.3–0.8 g/cm^3)

f (Hz)	Moisture content (%)										
	5	10	15	20	25	30	40	60	80	100	120
10^2	1.8	4	6	7	8	9	10	12	14	16	18
10^3	1.7	2.7	3.9	5.1	5.6	5.8	6.1	6.8	7.5	8.2	8.8
10^4	1.6	2	2.2	2.5	2.8	2.8	2.9	3	3.1	3.3	3.4
10^5	1.5	1.8	2.1	2.4	2.5	2.5	2.6	2.4	2.3	2.2	2.1
10^6	1.4	1.7	2.0	2.2	2.3	2.3	2.1	1.9	1.7	1.6	1.4
10^7	1.3	1.6	1.8	2.0	2.1	2.1	1.9	1.8	1.7	1.6	1.5
10^8	1.3	1.5	1.7	1.8	1.9	1.9	1.9	1.8	1.7	1.6	1.6
10^9	1.2	1.3	1.5	1.6	1.8	1.8	1.8	1.8	1.8	1.8	1.8
10^{10}	1.1	1.2	1.3	1.4	1.5	1.5	1.5[a]	1.5[a]	1.5[a]	1.5[a]	1.5[a]
10^{11} [a]	1.1	1.2	1.2	1.3	1.4	1.4	–	–	–	–	–

[a] Result of extrapolation

Table 5.6. Dependence of the dielectric constant of wood on the percentage of latewood in coniferous tree species at $f = 10^6$ Hz and t = 20 °C (Norimoto et al. 1978)

Latewood relative (%)	10			30		
W (%)	0	6.0	13.6	0	6.0	13.5
ρ_0 (g/cm^3)	0.43	0.42	0.41	0.58	0.57	0.56
ε'_L	2.24	3.11	5.17	2.71	4.90	6.81
ε'_R	1.75	2.13	2.92	1.96	2.75	3.36
ε'_T	1.66	1.95	2.53	2.07	2.96	3.63

The DC of the cell wall follows a parabolic curve with increasing moisture content up to the fiber saturation point (Fig. 5.8). At a frequency of 10^6 Hz, the value of $\varepsilon'_{\perp}$ (Uyemura 1960) increases from 4.8 (at W = 0%) to 34 (W = 28%). In accordance with data by Norimoto et al. (1978) the DC of a softwood cell wall increases from 4.9 to 11.6 with a transverse orientation of vector $\bar{E}$, at a frequency of 10^6 Hz and at a temperature of 20 °C when the moisture content increases from 0 to 13–14%. At the same time, the value of coefficient $k_1 = \varepsilon'_{\parallel}/\varepsilon'_{\perp}$ of the cell wall increases from 1.09 to 1.44.

In connection with the use of microwave techniques by industry, using the frequencies 0.915 and 2.375 GHz allowed for industrial applications, I conducted research on the dielectric properties of wood from tree species of industrial importance in Russia, using the above frequencies. The methods employed in these experiments are described in Sect. 3.3. The dielectric properties of wood

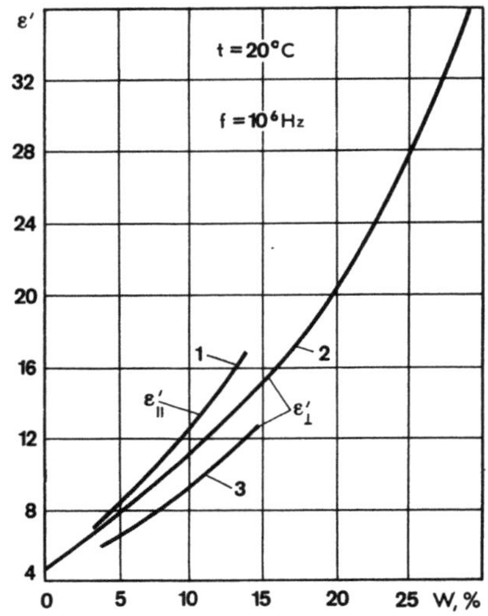

Fig. 5.8. Dielectric constant of the cell wall substance ε' as a function of the moisture content W up to 28%, at $t = 20\,°C$ and frequency 10^6 Hz, according to data of *1, 3* Norimoto et al. (1978); *2* Uyemura (1960)

were determined at moisture contents ranging from 0–120% and at temperatures from -30 to $+90\,°C$ with samples prepared from logs of spruce, larch, aspen, and birch. The samples (28–38 cm diameter) were sawed out of the central part of the trunks. The average density of samples of the oven-dry wood was as follows: spruce-0.42, larch – 0.63, birch – 0.60 and aspen – 0.47 g/cm^3. The ε' and $tg\delta$ values obtained are listed in Tables 5.7 and 5.8. As the difference between the values of ε' and $tg\delta$ at 70 and 90 °C was small, these values were combined in one column. The coefficient of variation of ε' varied from 1.2 to 13.5% that of $tg\delta$ changed within 2 – 20%, and the accuracy factors (index of the precision of investigation) were within 0.4–4.3 and 0.6–5.7%, respectively.

At a constant temperature, the value of ε' increases with increasing moisture content. At frequencies of 0.9×10^9–2.4×10^9 Hz, temperature changes from 20 °C to 90 °C do not influence the DC significantly. The difference between these values for all species at the extreme temperatures in this range does not exceed 15%. An increase in frequency decreases the DC value.

Experimental data relating to the values of the dielectric properties of spruce and birch wood measured at super high frequencies and at negative temperatures and at moisture contents from 41 to 62% are given in Table 5.9. The coefficient of variation of the DC values at negative temperatures and at frequencies of 0.915 and 2.375 GHz was 19%, and the accuracy factor was 3.6%. In the negative temperature range at these frequencies, the value of the DC to some extent increases with increasing temperature, but it is less subject to the influence of the moisture content than at positive temperatures (Fig. 5.9). At negative temperatures, the DC in the longitudinal direction is 1.5 times higher than in the transverse direction. An abrupt increase of the DC value is noted at a temperature near 0 °C. A further increase of the temperature affects the DC values only slightly since this increase lies within the error of the measurements.

Table 5.7. Dielectric constant (ε') and loss tangent ($tg\delta$) of spruce and larch at frequencies of 2.375 and 0.915 GHz

W	f	Spruce								Larch							
(%)	(GHz)	$t = 20\,°C$				$t = 70–90\,°C$				$t = 20\,°C$				$t = 70–90\,°C$			
		$\varepsilon'_\perp$	$tg\delta_\perp$	$\varepsilon'_\parallel$	$tg\delta_\parallel$	$\varepsilon'_\perp$	$tg\delta_\perp$	$\varepsilon'_\parallel$	$tg\delta_\parallel$	$\varepsilon'_\perp$	$tg\delta_\perp$	$\varepsilon'_\parallel$	$tg\delta_\parallel$	$\varepsilon'_\perp$	$tg\delta_\perp$	$\varepsilon'_\parallel$	$tg\delta_\parallel$
5	2.375	1.8	0.06	2.0	0.08	2.0	0.08	2.4	0.12	2.2	0.10	2.2	0.10	2.8	0.13	3.0	0.14
10		2.1	0.09	2.5	0.15	2.2	0.10	3.2	0.18	3.0	0.15	3.0	0.15	3.6	0.16	3.6	0.16
20		2.8	0.16	3.8	0.26	2.8	0.14	4.2	0.20	5.2	0.21	10.6	0.25	5.6	0.19	9.8	0.25
30		3.5	0.22	5.5	0.30	3.5	0.17	5.5	0.25	7.5	0.25	17.1	0.35	7.7	0.20	14.9	0.31
40		4.3	0.21	7.1	0.27	4.3	0.17	7.1	0.23	10.3	0.25	19.9	0.31	9.9	0.19	18.3	0.29
80		7.0	0.18	16	0.22	7.0	0.16	16	0.18	20	0.23	35	0.23	18	0.15	33	0.23
120		9.8	0.18	25	0.20	9.8	0.16	25	0.16	–	–	–	–	–	–	–	–
5	0.915	1.8	0.08	2.4	0.08	1.9	0.08	2.3	0.10	2.2	0.11	3.2	0.11	2.8	0.11	4.0	0.15
10		2.1	0.11	3.3	0.11	2.4	0.10	3.0	0.12	2.9	0.14	4.9	0.14	3.6	0.13	5.8	0.16
20		3.2	0.17	4.8	0.17	3.4	0.13	4.8	0.17	5.7	0.17	13.1	0.21	5.6	0.16	12.0	0.20
30		4.5	0.20	6.3	0.24	4.5	0.17	6.3	0.21	8.5	0.18	19.5	0.22	7.7	0.17	17.7	0.20
40		5.3	0.19	8.3	0.23	5.3	0.16	8.3	0.20	11.1	0.18	23	0.20	10.3	0.16	20	0.18
80		8.8	0.15	16.2	0.21	8.8	0.13	16.2	0.18	22	0.17	37	0.17	19.0	0.15	34	0.15
120		12.3	0.11	25	0.19	12.3	0.10	25	0.16	–	–	–	–	–	–	–	–

Table 5.8. Dielectric constant (ε') and loss tangent ($tg\delta$) of aspen and birch at frequencies of 2.375 and 0.915 GHz

W	f	Birch								Aspen							
(%)	(GHz)	$t = 20\,°C$				$t = 70–90\,°C$				$t = 20\,°C$				$t = 70–90\,°C$			
		$\varepsilon'_\perp$	$tg\delta_\perp$	$\varepsilon'_\parallel$	$tg\delta_\parallel$	$\varepsilon'_\perp$	$tg\delta_\perp$	$\varepsilon'_\parallel$	$tg\delta_\parallel$	$\varepsilon'_\perp$	$tg\delta_\perp$	$\varepsilon'_\parallel$	$tg\delta_\parallel$	$\varepsilon'_\perp$	$tg\delta_\perp$	$\varepsilon'_\parallel$	$tg\delta_\parallel$
5	2.375	2.1	0.06	2.3	0.09	2.3	0.08	3.1	0.10	1.9	0.08	1.9	0.10	2.0	0.07	2.9	0.11
10		2.5	0.09	2.7	0.14	2.7	0.10	3.7	0.13	2.2	0.11	2.6	0.17	2.2	0.08	3.8	0.16
20		3.2	0.16	4.2	0.26	3.3	0.13	4.9	0.19	2.9	0.19	4.9	0.27	2.9	0.11	5.5	0.23
30		4.1	0.22	6.7	0.28	4.1	0.14	6.7	0.22	3.7	0.23	7.3	0.36	3.7	0.15	7.3	0.27
40		5.1	0.22	11.1	0.26	5.1	0.14	11.1	0.22	4.6	0.22	11.4	0.34	4.6	0.15	10.2	0.25
60		7.0	0.21	21	0.25	7.0	0.14	21	0.22	6.2	0.21	17	0.29	6.2	0.15	15.2	0.21
80		9.0	0.21	27	0.25	9.0	0.14	27	0.20	7.8	0.19	23	0.23	7.8	0.14	20	0.18
5	0.915	2.3	0.08	2.5	0.10	2.6	0.08	3.4	0.10	2.0	0.07	2.2	0.09	2.2	0.07	2.5	0.11
10		2.8	0.10	3.5	0.14	3.1	0.10	4.7	0.12	2.5	0.09	2.7	0.13	2.6	0.08	3.2	0.14
20		3.9	0.16	6.7	0.20	4.1	0.14	7.5	0.14	3.3	0.16	4.9	0.20	3.4	0.12	5.5	0.18
30		5.2	0.20	10.8	0.21	5.2	0.14	10.8	0.16	4.1	0.22	8.1	0.24	4.1	0.18	8.1	0.18
40		6.7	0.18	14.1	0.20	6.7	0.14	14.1	0.14	5.1	0.21	11.4	0.23	5.1	0.17	10.5	0.18
60		10.1	0.15	21	0.17	10.1	0.12	21	0.14	7.6	0.18	17	0.22	6.7	0.15	15.2	0.18
80		13.4	0.13	27	0.15	13.4	0.10	27	0.14	10.2	0.15	23	0.21	10.2	0.13	20	0.18

Table 5.9. Dielectric properties of spruce and birch wood at negative temperatures, with $f = 2.375$ and at 0.915 GHz

t		− 5 °C				− 10 °C				− 15 °C				− 20 °C				− 30 °C			
Species	W (%)	$\varepsilon'_\perp$	$tg\delta_\perp$	$\varepsilon'_\parallel$	$tg\delta_\parallel$	$\varepsilon'_\perp$	$tg\delta_\perp$	$\varepsilon'_\parallel$	$tg\delta_\parallel$	$\varepsilon'_\perp$	$tg\delta_\perp$	$\varepsilon'_\parallel$	$tg\delta_\parallel$	$\varepsilon'_\perp$	$tg\delta_\perp$	$\varepsilon'_\parallel$	$tg\delta_\parallel$	$\varepsilon'_\perp$	$tg\delta_\perp$	$\varepsilon'_\parallel$	$tg\delta_\parallel$
$f = 2.375$ GHz																					
Spruce	41	2.6	0.24	4.0	0.28	2.5	0.22	3.7	0.26	2.3	0.2	3.5	0.24	2.2	0.2	3.2	0.24	1.9	0.2	2.9	0.24
	58	4.9	0.26	7.3	0.32	3.6	0.25	5.4	0.31	2.4	0.25	3.6	0.29	2.2	0.22	3.2	0.26	1.9	0.16	2.8	0.20
Birch	62	4.8	0.25	7.2	0.29	4.5	0.24	6.7	0.29	4.2	0.24	6.2	0.28	3.9	0.24	5.9	0.28	3.4	0.24	5.0	0.3
$f = 0.915$ GHz																					
Spruce	41	2.6	0.24	4.0	0.28	2.5	0.2	3.7	0.24	2.3	0.16	3.5	0.2	2.1	0.15	3.1	0.18	1.7	0.15	2.5	0.17
	58	2.8	0.22	4.2	0.26	2.6	0.18	3.8	0.22	2.4	0.15	3.6	0.19	2.3	0.15	3.5	0.17	2.1	0.13	3.1	0.15
Birch	62	4.2	0.26	6.4	0.32	4.0	0.25	6.0	0.29	3.8	0.23	5.6	0.27	3.5	0.23	5.3	0.28	3.1	0.24	4.7	0.28

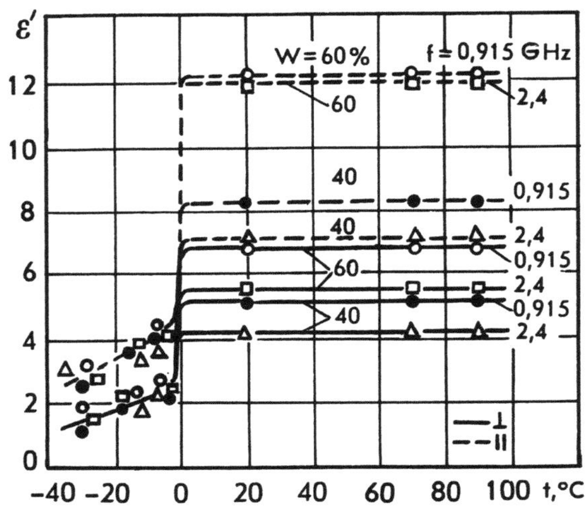

Fig. 5.9. Dependence of the dielectric constant of spruce wood on the temperature, at a moisture content $W = 40\%$ and 60%, as measured at microwave frequencies

The character of the interaction of a super high frequency electromagnetic field with the molecules of a substance differs from the character of their interaction at low frequencies, as the period of the field oscillation at microwave frequencies can be compared with the relaxation time of the molecules. A phase shift therefore arises between the field strength vectors and the polarization vector. This results in a reduction of ε' and in an increase of $tg\delta$ with increasing frequency. These results are valid for the majority of moist materials. In experiments with moist wood with an increase of frequency from 0.915 to 2.375 GHz, the value of ε' decreased in virtually all cases (Tables 5.7, 5.8).

At a moisture content above the fiber saturation point, the free water that fills the wood microcapillaries contain a small quantity of organic and mineral substances. Measurements of the dielectric properties of this moisture (extracted from newly felled trees) at frequencies of 0.915 and 2.375 GHz showed that they practically do not differ from those of pure water. Free moisture molecules interact with the microwave field independently of the cell wall substance and the bound moisture. Therefore, the change of ε' and $tg\delta$ of wood with the moisture content above the fiber saturation point is determined mainly by the dielectric properties of the free moisture and its relative volume. At negative temperatures, the wood consists of cell wall substance, air, ice, and nonfreezing, bound moisture. At microwave frequencies, the change of dielectric properties of the first three components with a change of temperature from -5 to $-30\,°C$ is small. Therefore, the quantity of nonfreezing bound moisture is the main factor determining the change of dielectric properties of moist wood at negative temperatures. However, the presence of supercooled, nonfreezing moisture in wood can significantly change the DC value of the wood.

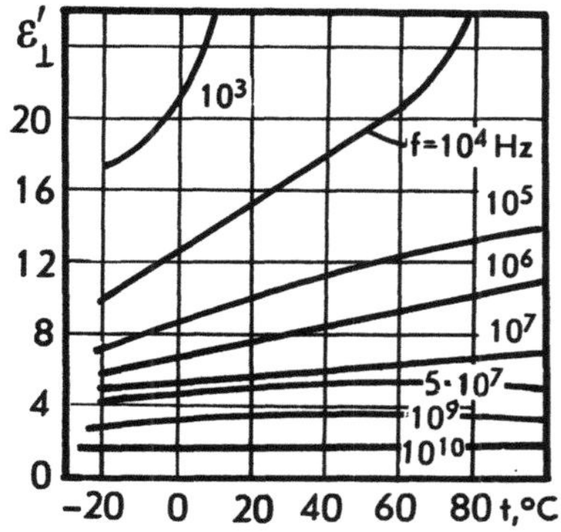

Fig. 5.10. Dependence of the dielectric constant $\varepsilon'_\perp$ of wood on temperature at frequencies from 10^3 to 10^{10} Hz, at moisture content $W = 25\%$, and density $\rho_0 = 0.4$–0.5 g/cm^3

With increasing temperature the DC of moist wood increases over the entire frequency range, i.e., the temperature coefficient of the DC $k^t_{\varepsilon'}$ is positive (Fig. 5.10). Depending upon the frequency, the moisture content, and the orientation of vector $\bar{E}$, the value of the temperature coefficient varies from zero to 0.14 K^{-1}. At frequencies up to 10^3 Hz and at a humidity level above 5–10%, the dependence of the DC on the temperature is nonlinear. This is the reeason why calculations of $\varepsilon' = \varphi(T)$ with the aid of temperature coefficients is difficult. Appendix 2 gives experimental data for the DC (James 1975) for temperatures equal to -20, $+25$, $+45$, and $+90$ °C and at wood moisture contents up to 100%.

At a moisture content up to the fiber saturation point, at frequencies 10^5 and higher, and in the temperature range from -70 up to 90 °C, the DC of wood increases continuously (at a constant moisture content) following characteristic curves, which are nearly linear.

At the moisture content above 20% with a rise of frequency up to 10^9 Hz, the value of $k^t_{\varepsilon'}$ decreases (Fig. 5.11), after which it rises again. Analysis of the temperature coefficients of DC has not established the influence of wood species or their density variations on their value; however, an essential influence of electric field orientation relative to the fiber direction has been noted. The average values of $k^t_{\varepsilon'_\perp}$ and $k^t_{\varepsilon'_\parallel}$ for the humidity range 5–30% are given in Table 5.10, and those for the range of 40–120% are given in Table 5.11.

At frequencies up to 10^9 Hz, the values of the temperature coefficient with vector $\bar{E}$ oriented parallel to the fiber direction are higher than those with the perpendicular orientation by 1.1–1.9 times. At frequencies of 10^9–10^{10} Hz, the values of $k^t_{\varepsilon'_\perp}$ and $k^t_{\varepsilon'_\parallel}$ are practically equal to each other at any moisture content.

At frequencies of 10^2–10^4 Hz, the DC values of ice in the temperature range from -3 to -30 °C are within 5–80. At these frequencies, the presence of ice essentially affects the DC value of wood with a moisture content above 25–30%. At frequencies above 10^4 Hz, the DC value of ice decreases to 3–4, and its influence on the DC value of wood is considerably lower. Therefore, the values of the temperature coefficients at negative temperatures in the wood humidity

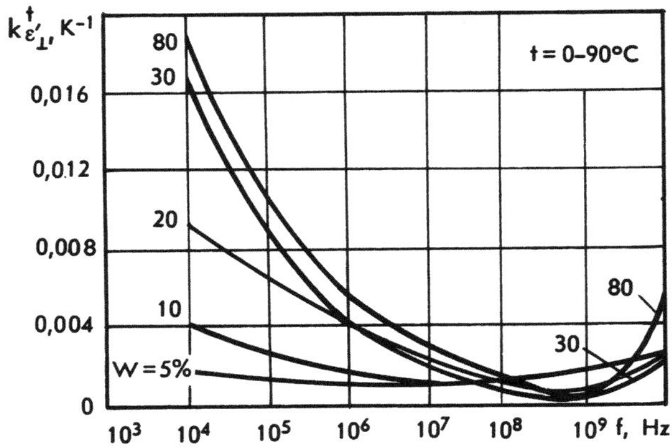

Fig. 5.11. Temperature coefficient $k^t_{\varepsilon'_\perp}$ of the dielectric constant of wood as a function of frequency *f* and moisture content *W* in the temperature range from 0 to 90 °C

Table 5.10. Average values of the temperature coefficients $k^t_{\varepsilon'} \times 10^3$ K^{-1} of wood with moisture content up to 30%

W (%)		5		10		15		20		25–30	
f (Hz)	*t* (°C)	⊥	‖	⊥	‖	⊥	‖	⊥	‖	⊥	‖
20	5–90	15	29								
10^2	5–90	10	19								
10^3	5–90	4	8	15	29						
10^4	− 30 to + 60	1.5	3	4	8	7	13	9	17	14	27
10^5	− 30 to + 90	1	1.9	3	6	4	8	5	10	–	–
10^6	− 30 to + 90	1	1.7	2	3.4	4	7	5	8.5	–	–
10^7	− 30 to + 90	1	1.5	1	1.5	1	1.5	2	3	–	–
10^8	− 30 to + 90	1	1.3	1	1.3	1	1.3	2	2.6	–	–
10^9	0–90	2	2	2	2	1.3	1.3	0.5	0.5	0	0
10^{10}	− 30 to + 90	2	2	2	2	2	2	2	2	–	–

range from 40 to 120% are practically identical. It should be noted that in this range of moisture content, the dependence $\varepsilon' = \varphi(T)$ is nonlinear, which fact is connected with the changes of the ice DC. However, the substitution of this characteristic with a linear one brings an error which lies within the variation limits of experimental data on the temperature coefficient values.

As is shown in Fig. 5.10, at a moisture content below the fiber saturation point, the curve of DC values does not evidence any abrupt changes at passing the point corresponding to 0 °C (Tiuri et al. 1980) as in this case the wood does not contain any free moisture. At a moisture content exceeding fiber saturation point, free moisture at temperatures about 0 °C turns from the solid state into

Table 5.11. Average values of the temperature coefficients $k^t_{\varepsilon'} \times 10^3\ \mathrm{K}^{-1}$ of wood with moisture content from 40 to 120%

t (°C)	0–90										−5 to −30	
W (%)	40		60		80		100		120		40–120	
f (Hz)	⊥	∥	⊥	∥	⊥	∥	⊥	∥	⊥	∥	⊥	∥
10^4	16	29	17	29	18	27	20	26	20	22	18	34
10^5	5	9	6	10	7	10.5	8	10.4	9	9.9	10	19
10^6	5	8	5	7.5	5	7	5	6	5	5.5	9	15.3
10^7	2	2.8	2	2.6	2	2.6	1	1.2	0	0	7	10.5
10^8	2	2.6	2	2.4	2	2.4	1	1.1	0	0	6	7.8
10^9	0	0	0	0	0	0	0	0	0	0	13	13
10^{10}	2	2	2	2	2	2	2	2	2	2	13	13

Table 5.12. Coefficient a' values for calculation of wood dielectric constant at $t = -20\,°\mathrm{C}$, using Eq. (5.4)

f (Hz)	Moisture content (%)						
	10	20	30	40	60	80	100
10^4	–	–	–	21	25	28	32
10^5	–	–	–	13.2	14.2	15.4	17
10^6	–	–	–	10.2	11.3	12.6	13.8
10^7	–	–	–	8.7	9.3	10.4	11.8
5×10^7–10^8	–	–	–	7.3	7.9	8.9	9.9
10^9–2.4×10^9	1.7	2.0	2.3	2.4	3.2	3.7	4.2
5.8×10^9	1.6	1.8	2.1	2.3	3.0	3.4	4.0
10^{10}	1.5	1.7	2.0	2.2	2.8	3.2	3.9

the liquid state, and its DC value sharply increases, which results in a sharp change of the wood DC value (Fig. 5.9).

For the determination of $\varepsilon'_{\perp(-20)}$ at a temperature $-20\,°\mathrm{C}$ as a function of wood density (at $\rho_0 = 0.3$–$0.8\ \mathrm{g/cm^3}$), the following equation can be used

$$\varepsilon'_{\perp(-20)} = a'\rho_0 + 1\ , \tag{5.4}$$

where

a' = empirical coefficient taken from Table 5.12

ρ_0 = density value of oven-dry wood ($\mathrm{g/cm^3}$).

The relationship of DC values in the longitudinal and transverse directions at $t = -20\,°\mathrm{C}$ and with a moisture content range 40–120% is according to experimental data from 1.5 to 2.1 depending on the frequency (Table 5.13).

Table 5.13. Average values of coefficient $k_1 = \varepsilon'_{\parallel}/\varepsilon'_{\perp}$ of wood at $t = -20\,°C$ and moisture content 40–120%

Frequency (Hz)	10^4	10^5	10^6	10^7	10^8	10^9	10^{10}
k_1	2.1	2.0	1.8	1.7	1.7	1.5	1.3[a]

[a] Result of extrapolation.

For determination of the DC value of wood with stipulated density and moisture content and at the required temperature in the frequency range from 10^4–10^{10} Hz, it is necessary to employ the following equation

$$\varepsilon' = \varepsilon'_{20}[1 + k^t_{\varepsilon'}(T - 293)] , \tag{5.5}$$

substituting the value of DC at 20 °C ($\varepsilon'_{\perp 20}$, $\varepsilon'_{\parallel 20}$) taken from Appendix 3 or calculated with the use of formulae, and the temperature coefficient $k^t_{\varepsilon'}$ taken from Tables 5.10 and 5.11.

For the determination of the DC value of frozen wood with moisture content 40–120% in the temperature range from -5 to -30 °C, the following relation can be used

$$\varepsilon' = \varepsilon'_{(-20)}[1 + k^t_{\varepsilon'}(T - 253)] , \tag{5.6}$$

where

$\varepsilon'_{(-20)}$ = the DC value at -20 °C ($\varepsilon'_{\perp(-20)}$ or $\varepsilon'_{\parallel(-20)}$) from Appendix 3 or calculated with the use of Eq. (5.4)

$k^t_{\varepsilon'}$ = temperature coefficient value in Table 5.11 ($t = -5$ to -30 °C).

An example: It is necessary to determine the DC value $\varepsilon'_{\perp}$ of wood with density in an oven-dry state equal to 0.5 g/cm^3 at a moisture content equal to 20%, at frequency 10^6 Hz, and with temperature -10 °C (263 K). The value of $\varepsilon'_{\perp}$ can be determined from Appendix 3, with reference to the following conditions: $f = 10^6$ Hz, $\rho_0 = 0.5$ g/cm^3, $W = 20\%$, and $t = +20$ °C. The value will be $\varepsilon'_{\perp} = 5.1$. From Table 5.10 we determine $k^t_{\varepsilon'_{\perp}} = 5 \times 10^{-3}\,K^{-1}$ and substitute this value into Eq. (5.5):

$$\varepsilon'_{\perp(-10)} = 5.1 \times [1 + 5 \times 10^{-3}(263 - 293)] = 4.34 .$$

5.2. The Dielectric Loss Tangent of Moist Wood

Dependence of the dielectric loss tangent on the tree species, wood density, moisture content, and frequency is more complicated that that for DC values. The loss tangent also possesses a bigger variation coefficient value. With measurements of $tg\delta$, the magnitude of errors is always bigger than with those of DC values. The differences between these data obtained by various authors are also more sizable. The values of $tg\delta$ at low frequencies (from 20 to 10^3 Hz) obtained by Mikhailovskaja (1972), Nanassy (1972), and James (1975) in their works differ from each other by tenfold and more. This fact emphasizes the complicacy of accurate measurements of $tg\delta$ at these frequencies. It is impossible to give preference to any of these data.

According to Mikhailovskaja (1972), at a frequency of 50 Hz and with a rise of moisture content to 30%, the value of larch wood $tg\delta$ increses to 2.2–2.5, and then the rate of increase becomes less; at $W = 70\%$ the $tg\delta$ value reaches 2.6–3.3, while the value of $tg\delta_{\parallel}$ exceeds the value of $tg\delta_{\perp}$ (Fig. 5.12). According to data by James (1975), at a frequency of 20 Hz and with an increase of moisture content to 30%, the value of softwood $tg\delta$ rapidly increases to 23–38, and with a further increase of moisture content, it rises more slowly, up to the value of 26–45 at $W = 120\%$ (Fig. 5.12). Of a similar character is the dependence of the $tg\delta$ value of oak wood. At a moisture content above 5% according to James (1975), $tg\delta_{\parallel} < tg\delta_{\perp}$, which is difficult to explain.

With a rise of frequency starting from 20 Hz and at a moisture content up to 15%, the value of $tg\delta$ diminishes (Fig. 5.13). With a rise of humidity, the value of $tg\delta$ increases and passes its maximum in the frequency range 10^2–10^3 Hz and

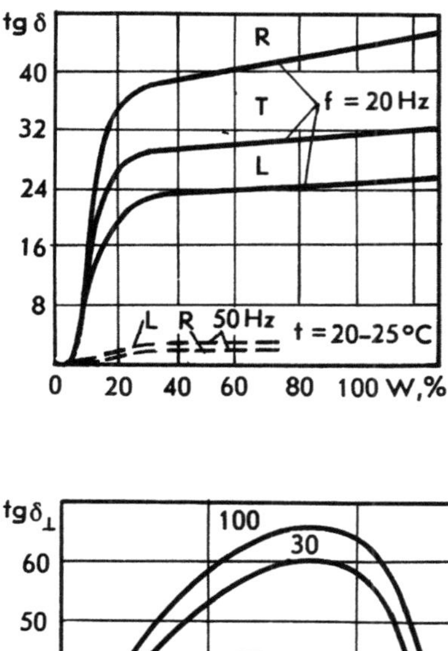

Fig. 5.12. Loss tangent value, $tg\,\delta$, of moist wood at low frequencies in the longitudinal (L), radial (R), and tangential (T) directions. *Dotted lines* correspond to larch wood values of $tg\,\delta$ (Mikhailovskaja 1972); *solid curves* correspond to $tg\,\delta$ values of Douglas-fir wood (James 1975)

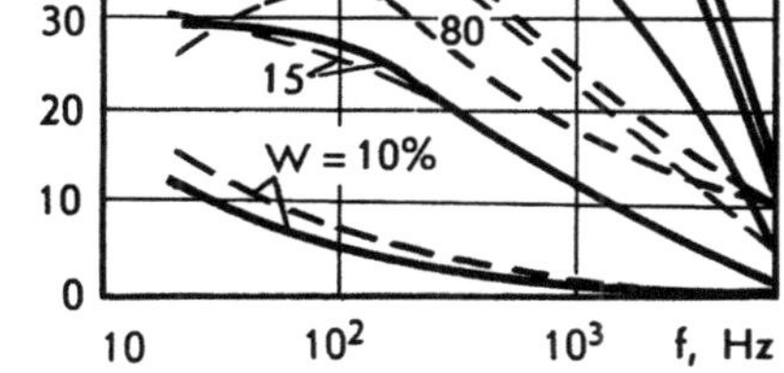

Fig. 5.13. Dependence of wood loss tangent values $tg\,\delta_{\perp}$ on moisture content W and frequency f in the range from 20 to 10^4 Hz at $t = 25\,°C$ (James 1975). *Solid lines* refer to softwood; *dotted lines* are for white oak ($tg\,\delta_R$)

then rapidly decreases. With an increase of moisture content above 30%, the value of *tgδ* of softwood increases and that of oak wood diminishes. It is difficult to ascertain the influence of wood density on the *tgδ* value in the low frequency range because of a lack of data. In most cases, the experimental data state that $tg\delta_R$ is higher than $tg\delta_T$.

At frequencies of 10^4–10^{10} Hz, the main influences on the *tgδ* value are the wood moisture content and density values. In the frequency range from 10^4 to 10^{10} Hz with a rise of humidity to 30%, the value of *tgδ* rapidly increases (Fig. 5.14), and then, depending on frequency, it grows more slowly (10^8, 10^{10} Hz), and at frequencies of 10^6 and 10^7 Hz, its value is approximately the same, or drops (at 10^9 Hz). In this frequency range, at a moisture content up to 10–12%, and with a rise of frequency starting from 10^5 Hz, the value of *tgδ* increases (Fig. 5.15), reaches its maximum in the range 10^8–10^9 Hz, and then drops. At the moisture content exceeding 15%, the value of *tgδ* with a rise of frequency starting from 10^4 Hz first decreases, passes its minimum in the range 10^7–10^8 Hz, and after that increases. At frequencies of 10^4–10^{10} Hz, an increase of moisture content above 30% only slightly influences the value of *tgδ*.

At a moisture content exceeding 30%, the greatest influence on the wood *tgδ* value is rendered by the quantity of free water and its *tgδ* value. The minimum value of wood *tgδ* at frequencies of 10^7–10^8 Hz is stipulated by the minimum value of the free water *tgδ* at these frequencies. With a rise of frequency from 5×10^7 Hz, the value of *tgδ* of water increases, reaching its maximum (at 20 °C) at frequencies near 60 GHz (Rosenberg 1979). This is the reason why the maximum value of the wood loss tangent at $t = 20\,°C$ with a moisture content above the fiber saturation point can also be expected at frequencies near 60 GHz. However, this should be experimentally verified. Hearmon and Burcham (1954) in their research established that a correlation exists between the moist wood density and *tgδ* value. With an increase of frequency, the correlation bond's magnitude grows, and with an increase in moisture content, it

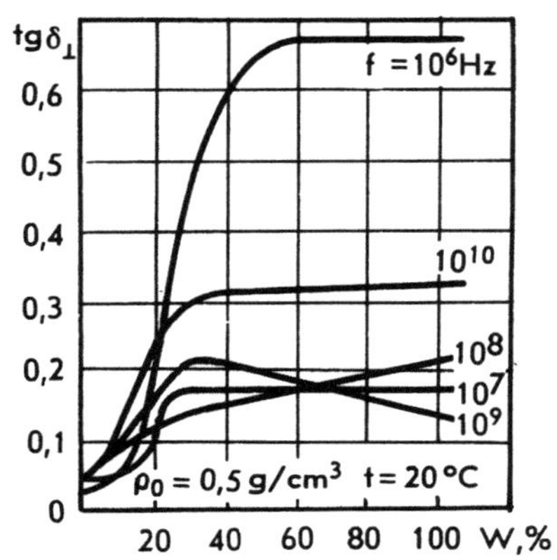

Fig. 5.14. Loss tangent value $tg\delta_\perp$ of moist wood as a function of moisture content, at frequencies from 10^6 to 10^{10} Hz, and at temperature $t = 20\,°C$: *f* frequency; *W* moisture content

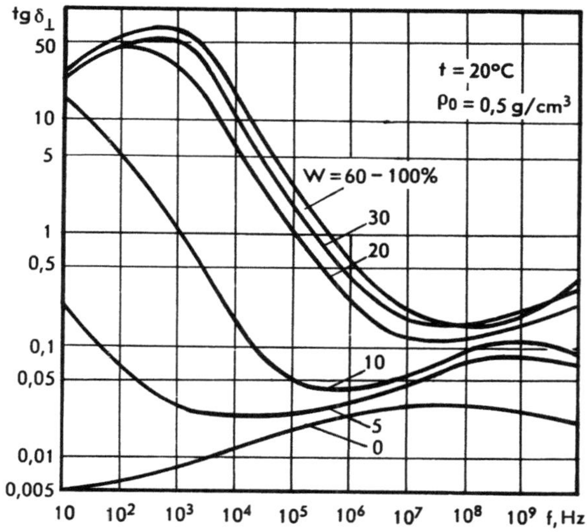

Fig. 5.15. Frequency response of the loss tangent values of wood at $t = 20\,°C$: f frequency; W moisture content; ρ_0 density of oven-dry wood

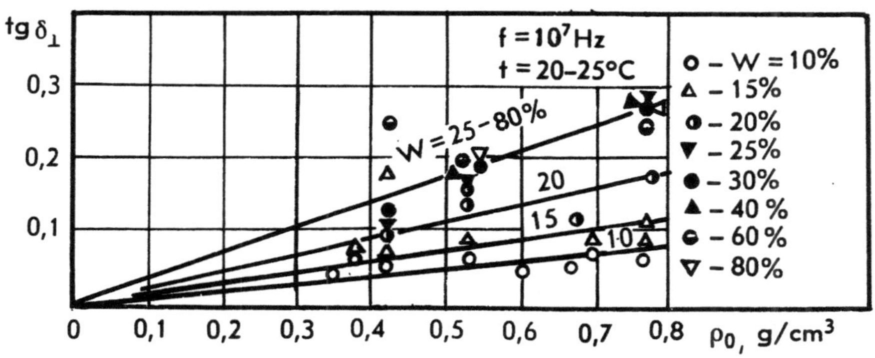

Fig. 5.16. Dependence of loss tangent values $tg\delta_\perp$ of moist wood on the density value ρ_0 of oven-dry wood: f frequency; W moisture content

diminishes. The correlation coefficient for moist wood with vector $\bar{E}$ oriented in the R, L, and T directions varied within the limits of 0.36–0.88. Analysis of the experimental data showed that at frequencies above 10^4 Hz the dependence of the $tg\delta$ values on the wood density in the moisture content range from 0 to 100% with the vector $\bar{E}$ orientation perpendicular to the fibers can be represented by straight lines passing through the origin of the coordinates (the point of $\rho_0 = 0$, $tg\delta = 0$; Fig. 5.16). At a certain moisture content value, the dependence

Table 5.14. Coefficient b value in Eq. (5.7) ($\rho_0 = 0.3$–0.8 g/cm^3, $t = 20$–$25\,^\circ$C)

f (Hz)	Moisture content (%)									
	5	10	15	20	25	30	40	60	80	100
10^4	0.05	0.3	3	11	20	20	20	20	20	20
10^5	0.05	0.09	0.6	1.9	3.0	3.0	3.5	4.1	4.1	4.1
10^6	0.07	0.09	0.2	0.48	0.92	0.92	1.12	1.34	1.34	1.34
10^7	0.07	0.10	0.14	0.22	0.34	0.34	0.34	0.34	0.34	0.34
10^8	0.15	0.2	0.22	0.24	0.29	0.29	0.29	0.36	0.40	0.40
10^9	0.15	0.2	0.26	0.31	0.35	0.43	–	–	–	–
2.4×10^9	0.15	0.21	0.28	0.35	0.39	0.44	–	–	–	–
10^{10}	0.14	0.17	0.32	0.47	0.57	0.62	–	–	–	–

of $tg\delta$ on the wood density (within the limits of $\rho_0 = 0.3$–0.8 g/cm^3) at frequencies of 10^4–10^{10} Hz, and at a temperature of 20–25 °C can be expressed by the following empirical relation

$$tg\delta_\perp = b\rho_0\,, \tag{5.7}$$

where

ρ_0 = oven-dry wood density (g/cm^3)
b = coefficient whose values depend on frequency and moisture content (Table 5.14).

The deviations of experimental data relating to $tg\delta_\perp$ values obtained by most investigators from the average values calculated with the use of Eq. (5.7) at frequencies equal to 10^4 and 10^6 Hz do not exceed $\pm$ 40%; at frequencies of 10^5, 10^7, and 10^8 Hz, these deviations are within $\pm$ 35%; and at frequencies of 10^9–10^{10} Hz, they are $\pm$ 30%. Taking into consideration that the value of $tg\delta$ of moist wood has a large coefficient of variation and that the errors of measurements of $tg\delta$ are considerable (especially for wood with a high moisture content), such differences between the calculated and experimental data can be accepted as satisfactory. It should be noted that at frequencies of 10^4–10^6 Hz, the difference between the values of $tg\delta_\perp$ of moist wood obtained by various investigators under similar conditions in some cases is as high as 1.5–2 times.

The values of $tg\delta_\perp$ of wood with a moisture content above 30%, at frequencies of 10^9–10^{10} Hz, and at $t = 20$–$25\,^\circ$C can be calculated using Eq. (2.5) with the empirical correction factor "c_1" introduced into this equation

$$tg\delta_\perp = (tg\delta_{p\perp} + V_4\,tg\delta_w)/c_1\,, \tag{5.8}$$

where

$tg\delta_{p\perp}$ = loss tangent of wood with a moisture content equal to 30% determined from Eq. (5.7)
$tg\delta_w$ = loss tangent of water (Table 2.8)

Table 5.15. Values for correction factor c_1 in Eq. (5.8)

Frequency (Hz)	Moisture content (%)			
	40	60	80	100
10^9	1	1.1	1.2	1.4
$2.4 \times 10^9 - 10^{10}$	1.05	1.2	1.3	1.4

c_1 = empirical correction factor whose values are given in Table 5.15
V_4 = relative volume of water in the wood (Table 2.1).

Values for the loss tangent of moist wood with an orientation of vector $\bar{E}$ perpendicular to the fiber direction, in the frequency range from 10^4 to 10^{10} Hz, and at different temperatures are given in Appendix 3.

The difference between $tg\delta_R$ and $tg\delta_T$ values at a moisture content of 5–10% is insignificant, but in most cases, $tg\delta_R > tg\delta_T$. With a rise of moisture content, the relation $k_{2(R/T)} = tg\delta_R / tg\delta_T$ increases, and at frequencies of 10^4–10^8 Hz, it is within 1–1.7 and is determined by the volume of rays. Oak wood has a higher $k_{2(R/T)}$ value than wood species with a ray volume less than 12%. At frequencies of 10^9–10^{10} Hz, the efforts to valuate the difference between $tg\delta_R$ and $tg\delta_T$ values have not been successful. The values of coefficient $k_{2(R/T)}$ are given in Table 5.16. The difference between $tg\delta_R$ and $tg\delta_T$ decreases with increasing frequency. If the value of $tg\delta_\perp$ is known, the values of $tg\delta_R$ and $tg\delta_T$ can be calculated with the aid of Eq. (4.6), using the coefficients from Table 5.16.

With the orientation of vector $\bar{E}$ parallel to the fiber direction and at low frequencies up to 10^3 Hz, the value of $tg\delta_\parallel$ proves to be higher or lower than $tg\delta_\perp$ (Mikhailovskaja 1972, James 1975). At frequencies of 10^4 Hz and higher, the relation of $tg\delta_\parallel / tg\delta_\perp$ is virtually the same for coniferous and angiosperm species. The value of $k_2 = tg\delta_\parallel / tg\delta_\perp$ for moist wood at a temperature of 20–25 °C is presented in Table 5.17. The value of k_2 for wood with a moisture content from 5 to 120% and at frequencies of 10^4–10^{10} Hz varies from 1.2 to 4.2. At frequencies of 10^4–10^7 Hz, the value of k_2 increases with increasing moisture content, but at higher frequencies it diminishes. For the determination of $tg\delta_\parallel$ under specified conditions, it is necessary to multiply the value of $tg\delta_\perp$ taken from Appendix 3 by the corresponding coefficient k_2 from Table 5.17.

The loss tangent value of moist wood has a complicated dependence on temperature (Fig. 5.17). At frequencies of 20–10^3 Hz, the change of $tg\delta$ with temperature has a dependence curve approaching a parabolic shape. The average values of the temperature coefficient in the range of 20–90 °C change depending on the moisture content as follows: at frequency 20 Hz, within 3×10^{-3} and 200×10^{-3}; at $f = 10^2$ Hz, from 3×10^{-3} to 120×10^{-3}; and at $f = 10^3$ Hz, from 7×10^{-3} to 100×10^{-3} K^{-1}.

At frequencies of 10^4–10^8 Hz and in the temperature range 20–90 °C $tg\delta$ changes with increasing temperature following curves that are nearly linear. The

Table 5.16. Coefficient $k_{2(R/T)} = tg\delta_R/tg\delta_T$ values for wood at $t = 20$–$25\,°C$

f (Hz)	Species	Moisture content (%)					
		5	10	15	20	25–30	40–120
10^4–10^6	Wood species with a ray volume up to 12%	1	1.05–1.1	1.1–1.2	1.2–1.3	1.3–1.4	1.35–1.55
	Oak and wood species with a ray volume exceeding 18%	1–1.1	1.1–1.2	1.2–1.35	1.25–1.4	1.3–1.45	1.35–1.7
10^7–10^8	Wood species with a ray volume up to 12%	1	1	1–1.1	1.05–1.1	1.05–1.1	1.05–1.1
	Oak and wood species with a ray volume exceeding 18%	1–1.1	1.1–1.15	1.1–1.25	1.2–1.3	1.2–1.35	1.05–1.3
10^9–10^{10}	All wood species	1	1	1	1	1	1

Table 5.17. Average values of coefficient $k_2 = tg\delta_{\parallel}/tg\delta_{\perp}$ at $t = 20$–$25\,°C$

f (Hz)	Moisture content (%)										
	5	10	15	20	25	30	40	60	80	100	120
10^4	1.6	1.6	1.6	1.6	1.6	1.6	1.8	2.2	2.5	2.8	3.2
10^5	1.3	1.4	1.6	1.7	1.7	1.8	2	2.6	3.2	3.7	4.2
10^6	1.4	1.6	1.6	1.6	1.6	1.6	1.6	1.7	1.8	1.9	2.0
10^7	1.4	1.4	1.4	1.4	1.4	1.4	1.4	1.5	1.5	1.6	1.6
10^8	1.4	1.4	1.3	1.3	1.3	1.3	1.3	1.3	1.3	1.3	1.3
10^9	1.3	1.3	1.2	1.2	1.2	1.2	1.2	1.2	1.2	1.2	1.2
2.4×10^9	1.5	1.5	1.5	1.4	1.4	1.3	1.2	1.2	1.2	1.2	1.2
10^{10}	2.0	1.8	1.6	1.6	1.6	1.5	1.4[a]	1.4[a]	1.4[a]	1.4[a]	1.4[a]

[a] Result of extrapolation.

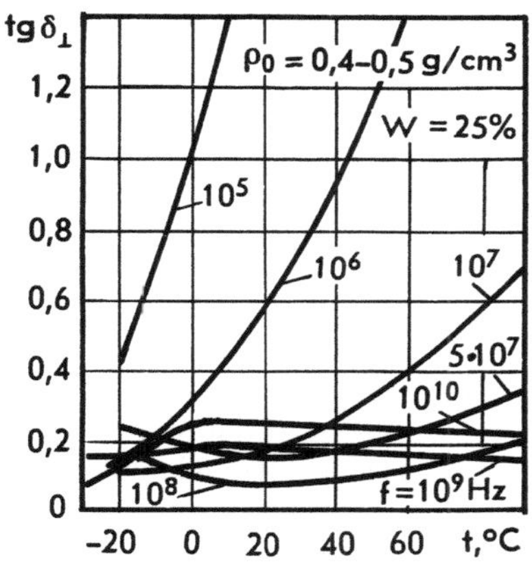

Fig. 5.17. Dependence of the loss tangent value $tg\delta_{\perp}$ of wood with a moisture content $W = 25\%$ on temperature t, at the density value $\rho_0 = 0.4$–$0.5\ g/cm^3$, and at frequencies f from 10^5 to 10^{10} Hz

value of $tg\delta$ under the specified conditions can be calculated with the aid of temperature coefficients using the formula

$$tg\delta = tg\delta_{20}\,[1 + k^t_{tg\delta}(T - 293)]\,, \tag{5.9}$$

where

$tg\delta_{20}$ = value of loss tangent at 20 °C

$k^t_{tg\delta}$ = temperature coefficient of loss tangent in the specified temperature range.

Depending on the conditions, the temperature coefficient varies as follows: at frequencies of 10^4–10^5 Hz and at moisture content of 5–20%, it changes from -6×10^{-3} to $+113 \times 10^{-3}\ K^{-1}$, and at a moisture content above 20%, it is equal to 10×10^{-3} to $60 \times 10^{-3}\ K^{-1}$; at frequencies of 10^6–10^8 Hz and at a moisture content of 5%–20%, it changes from -13×10^{-3} to $+40 \times 10^{-3}$ K^{-1}, and at a moisture content above 20%, it is equal to -15×10^{-3} to

Table 5.18. Average values of the temperature coefficient of the wood loss tangent $k^t_{tg\delta} \times 10^3\,K^{-1}$

Frequency (Hz)	Temperature (°C)	Moisture content (%)						
		5	10	20	30	40	80	100
10^4	20–90	70	80	50	40	35	25	20
	−5 to −20	−5	50	80	80	70	50	40
10^5	20–90	0	70	55	50	45	30	25
	−5 to −20	−10	10	55	55	50	45	40
10^6	20–90	−8	30	30	30	30	32	33
	−5 to −20	−13	−5	35	40	40	30	20
10^7	20–90	−6	10	17	24	25	33	37
	−5 to −20	−7	0	15	30	30	20	15
10^8	20–90	−3	−5	15	23	23	24	25
	−5 to −20	−1	−12	−12	−12	−14	−18	−20
10^9	20–90	1.5	0	−2	−2.5	−2.5	−2	−1.8
	−5 to −20	6	9	13	17	17	17	17
	−20 to −30	–	–	–	10	10	10	10
2.4×10^9	20–90	3.5	0	−2.7	−3.1	−3.1	−3.1	−3.2
	−5 to −20	6	9	13	17	17	17	17
	−20 to −30	–	–	–	10	10	10	10
10^{10}	0–30	12	17	28	28			
	0 to −20	8	10	15				

$+ 60 \times 10^{-3}\,K^{-1}$. Any real difference between the temperature coefficients of $tg\delta$ in the longitudinal and transverse directions has not been established, and therefore the coefficients given in Table 5.18 can be used for the calculation of both $tg\delta_\perp$ and $tg\delta_{||}$.

Figure 5.18 shows the dependence of $k^t_{tg\delta}$ on the frequency in the temperature range 20–90 °C. At a wood moisture content equal to 20–30%, the coefficient $k^t_{tg\delta}$ has a maximum at a frequency of 10^5 Hz and a minimum value at a frequency near 3×10^9 Hz. At a moisture content of 5%, the minimum value of $k^t_{tg\delta}$ is at a frequency of 10^6 Hz, and at 10% it is at a frequency of 10^8 Hz.

At frequencies of 10^9–2.4×10^9 Hz, in the temperature range of 20–90 °C, and at a moisture content of up to 10%, the value of $k^t_{tg\delta}$ is positive, but at a moisture content above 10%, it is negative. My own research showed that this coefficient, at a moisture content above 10%, varies from -1.3×10^{-3} to $-6.2 \times 10^{-3}\,K^{-1}$. At frequencies near 10^{10} Hz, in the temperature range from 0 to 30%, and with a moisture content increasing from 5 to 30%, the value of $k^t_{tg\delta}$ increases from 12×10^{-3} to $28 \times 10^{-3}\,K^{-1}$.

Tables 5.7 and 5.8 contain values of $tg\delta_\perp$ and $tg\delta_{||}$ for spruce, larch, birch, and aspen wood obtained in experiments carried out at industrially allowed frequencies of 0.915 and 2.375 GHz. At these frequencies, the value of $tg\delta$ increases with an increase in moisture content from zero to the fiber saturation point, but with a further increase of the moisture content, it decreases. These

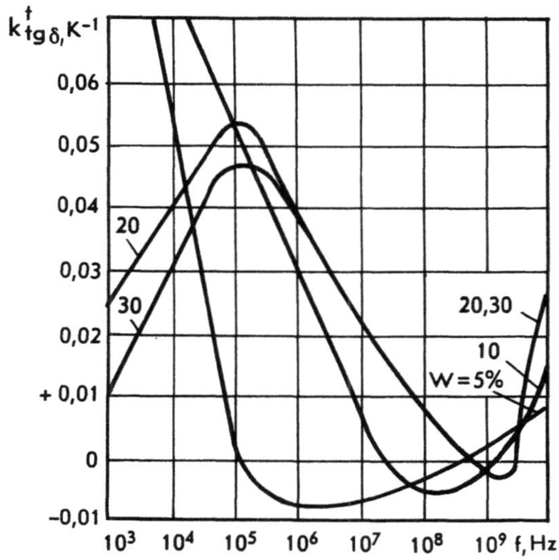

Fig. 5.18. Frequency response of the temperature coefficient $k^t_{tg\delta}$ of the wood loss tangent in the temperature range 20–90 °C at a moisture content $W = 5$–30%

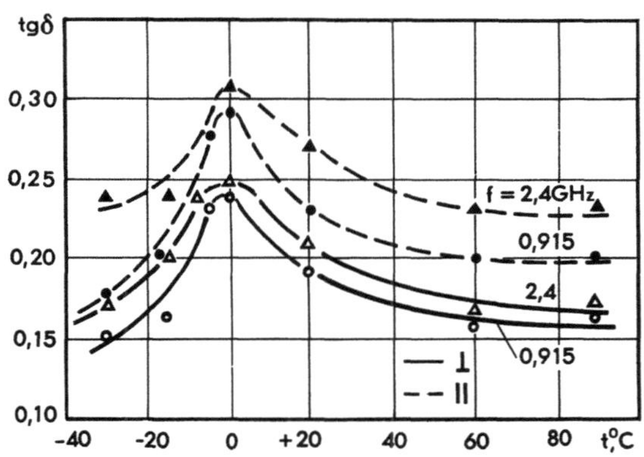

Fig. 5.19. Temperature dependence of $tg\delta$ of spruce wood at frequencies of 0.915 and 2.4 GHz and at a moisture of 40%

correlations are the same for all wood species. With an increase in temperature starting from −30 °C, the value of $tg\delta$ increases (Fig. 5.19), passes its maximum at 0 °C, and then decreases. With increasing field frequency, the value of $tg\delta$ increases. At negative temperatures, the coefficient of variation of $tg\delta$ was equal to 27%, which is considerably higher than at positive temperatures. The quantity of nonfreezing, bound water in the wood has a certain influence over the value of $tg\delta$. Depending on the conditions of freezing or defrosting, the quantity of the nonfreezing moisture in the wood at identical temperatures is subject to changes, which leads to a great variability of $tg\delta$.

At a temperature of $-20\,°C$ and in the frequency range 10^4–10^{10} Hz, the dependence of $tg\delta$ on the density of the wood is expressed by the following empirical formula

$$tg\delta_\perp = b'\rho_0 \,, \tag{5.10}$$

where

b' = coefficient whose value depends on frequency and moisture content, as shown in Table 5.19

ρ_0 = density value of oven-dry wood (g/cm^3).

At temperatures from -5 to $-30\,°C$, the dependence of $tg\delta$ on temperature is nearly linear, and it becomes possible to determine $tg\delta$ in this region with the aid of temperature coefficients, using the formula

$$tg\delta = tg\delta_{(-20)}[1 + \kappa^t_{tg\delta}(T - 253)] \,, \tag{5.11}$$

where

$tg\delta_{(-20)}$ = loss tangent value at $t = -20\,°C$, calculated from Eq. (5.10)

$\kappa^t_{tg\delta}$ = temperature coefficient for negative temperatures, the value of which can be taken from Table 5.18.

Table 5.19. Values of coeffieient b' in Eq. (5.10) ($\rho_0 = 0.3$–0.8 g/cm^3, $t = -20\,°C$)

f (Hz)	Moisture content (%)				
	10	20	30	60	100
10^4	0.14	1.0	2.0	2.5	3.5
10^5	0.11	0.44	0.7	0.9	1.2
10^6	0.13	0.24	0.3	0.3	0.27
10^7	0.21	0.2	0.2	0.18	0.15
10^8	0.26	0.4	0.4	0.4	0.33
10^9–2.4×10^9	0.15	0.28	0.4	0.4	0.4
5.8×10^9	0.13	0.23	0.33	0.33	0.33
10^{10}	0.10	0.18	0.26	0.26	0.26

Table 5.20. Average values of coefficient $k_2 = tg\delta_{\parallel}/tg\delta_\perp$ for wood at $t = -20\,°C$

Moisture content (%)	Frequency (Hz)						
	10^3	10^4	10^5	10^6	10^7	10^8	10^9–2.4×10^9
10–25	1.4	1.4	1.4	1.4	1.4	1.4	–
30–120	1.2	1.3	1.3	1.4	1.4	1.3	1.2

At a temperature equal to $-20\,°C$, the relation $k_2 = tg\delta_{\|}/tg\delta_{\perp}$ at frequencies of 10^3 Hz and higher averages 1.2–1.4 (Table 5.20).

For the determination of the $tg\delta$ value of wood with a specified density and moisture content at positive temperatures and in the frequency range of 10^4–10^{10} Hz, the values of $tg\delta_{\perp 20}$ and $tg\delta_{\| 20}$ (taken from Appendix 3 or calculated) and of the temperature coefficient $k^t_{tg\delta}$ (from Table 5.18) should be substituted into Eq. (5.9). For the determination of the $tg\delta_{\perp}$ value of frozen wood, the loss tangent value at $t = -20\,°C$ [$tg\delta_{\perp(-20)}$ or $tg\delta_{\|(-20)}$] and corresponding temperature coefficient should be substituted into Eq. (5.11).

An example: Let us determine the value of $tg\delta$ of wood with a density in the oven-dry state of $\rho_0 = 0.5\,g/cm^3$ at a moisture content equal to 30%, at a frequency of 10^9 Hz, at a temperature of $-10\,°C$ (263 K), and with vector $\bar{E}$ oriented parallel to the fiber direction. At first, the value of $tg\delta_{\perp(-20)}$ should be found using Eq. (5.10) with $b' = 0.4$ (taken from Table 5.19):

$$tg\delta_{(-20)} = b'\rho_0 = 0.4 \times 0.5 = 0.2 .$$

From Table 5.20, we find that

$$k_2 = tg\delta_{\|}/tg\delta_{\perp} = 1.2 .$$

Finally, calculate as follows

$$tg\delta_{\|(-20)} = tg\delta_{\perp(-20)} \times k_2 = 0.2 \times 1.2 = 0.24 .$$

Using Eq. (5.11) and substituting the temperature coefficient (from Table 5.18), we find the required value of the loss tangent at $t = -10\,°C$ ($k^t_{tg\delta} = 0.017$):

$$\begin{aligned} tg\delta_{\|(-10)} &= tg\delta_{\|(-20)}[1 + k^t_{tg\delta}(T - 253)] \\ &= 0.24[1 + 0.017(263 - 253)] = 0.28 . \end{aligned}$$

5.3 The Dielectric Loss Factor of Moist Wood

The loss factor ε'' characterizes the dissipation of electromagnetic field energy in the wood; the influence of moisture content on the loss factor proves to be decisive.

At a moisture content of up to 10%, at frequencies above 10^4 Hz (or at a moisture content below 5%–6% over the entire frequency range), and at $t = 20\,°C$, the value of $\varepsilon''_{\perp}$ is less than 1 and has its peak at a frequency near 10^8 Hz (Fig. 5.20). At the same time, the point of maximum value of $\varepsilon''_{\perp}$ is shifted to the lower frequency region in comparison with the point of maximum value of $tg\delta_{\perp}$.

At a moisture content above 20%, at frequencies of 20–100 Hz, and at $t = 20\,°C$, the loss factor value is as high as 10^4–10^7. With increasing frequency, the value of $\varepsilon''_{\perp}$ diminishes, passing its minimum at a moisture content of 20–30% and at a frequency 10^7–10^8 Hz, the minimum value of $\varepsilon''_{\perp}$ being about

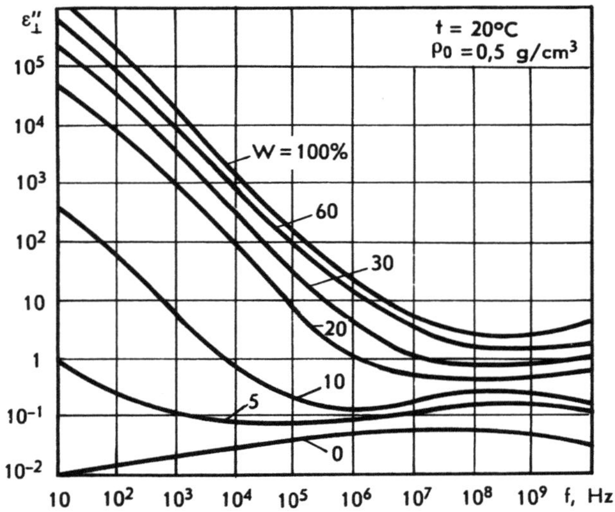

Fig. 5.20. Frequency response of the wood dielectric loss factor $\varepsilon''_{\perp}$ value at 20 °C: W moisture content; ρ_0 density of oven-dry wood

0.5–0.8. With a further increase in frequency, the value of the loss factor gradually increases. At a moisture content of 60–100%, the minimum of $\varepsilon''_{\perp}$ shifts in the direction of higher frequencies, its value being within the limits of 1.5–2.3.

At a moisture content above 30%, the loss factor value is profoundly affected by the quantity of free water and by its dielectric properties. The maximum value of the loss factor of water at $t = 20$ °C is approximately at a frequency of 15×10^9 Hz (Rosenberg 1979). Therefore, the maximum value of ε'' for wood with a moisture content above 30% can be expected at a frequency near 15×10^9 Hz.

The high loss factors of moist wood at low frequencies can be explained by a high ionic conductivity of moist wood. With increasing frequency, the influence of this factor decreases, and the relaxation losses begin to play a decisive role. With a change of wood density, temperature, and the orientation of vector $\bar{E}$ relative to the fiber direction, the values of the loss factor change, but the general character of its frequency response remains similar to that plotted in Fig. 5.20.

An increase in wood density increases the loss factor value under all conditions. The anisotropy of wood affects the loss factor value more than it does the values of the DC and loss tangent. At a temperature of 20–25 °C, the coefficient $k_3 = \varepsilon''_{\parallel}/\varepsilon''_{\perp}$ behaves differently at various frequencies. With an increase of moisture content from 5 to 100% and at frequencies of 10^4–10^5 Hz, the k_3 value increases from 2–2.6 to 8–9, and at frequencies of 10^6–10^{10} Hz it has its maximum at a moisture content of 25–30%. Average values for coefficient k_3 are given in Table 5.21.

Table 5.21. Average values of coefficient $k_3 = \varepsilon''_{\parallel}/\varepsilon''_{\perp}$ of moist wood at frequencies of 10^4–10^{10} Hz ($t = 20$–$25\,°C$, $\rho_0 = 0.3$–$0.8\ g/cm^3$)

f (Hz)	Moisture content (%)					
	5	10	20	30	60	100
10^4	2.6	3.2	4.0	4.5	6.6	9.2
10^5	2.0	2.5	4.0	4.5	6.2	8.1
10^6	2.0	2.7	3.5	3.7	3.2	3.0
10^7	1.8	2.2	2.8	2.9	2.7	2.6
10^8	1.8	2.1	2.3	2.5	2.3	2.1
10^9	1.6	1.7	1.9	2.2	2.2	2.2
10^{10}	2.2	2.2	2.2	2.3	2.1	2.1

The influence of temperature on ε'' of moist wood is complicated in character. According to Handa et al. (1982), at a temperature equal to $-196\,°C$, the loss factor of wood with a moisture content of 16% and in a frequency range from 30 to 10^6 Hz remains virtually unchanged, and its value is equal to about 0.005. With increasing temperature in the negative range, the value of ε'' increases at first and then has maximum values at every frequency (from 30 to 10^6 Hz) in a temperature range from -120 to $-40\,°C$. At a moisture content below the fiber saturation point, the loss factor increases also at positive temperatures. At a moisture content above the fiber saturation point, while crossing the temperature point equal to $0\,°C$, the loss factor coefficient undergoes a significant change (Fig. 5.21) which is associated with the transfer of ice, which had been formed from free moisture, from the solid into the liquid state. With an increase in temperature above $0\,°C$, the value of ε'' can increase or decrease depending on the quantity of free moisture in the wood and on its properties, which change considerably with any change in frequency (see Fig. 2.7).

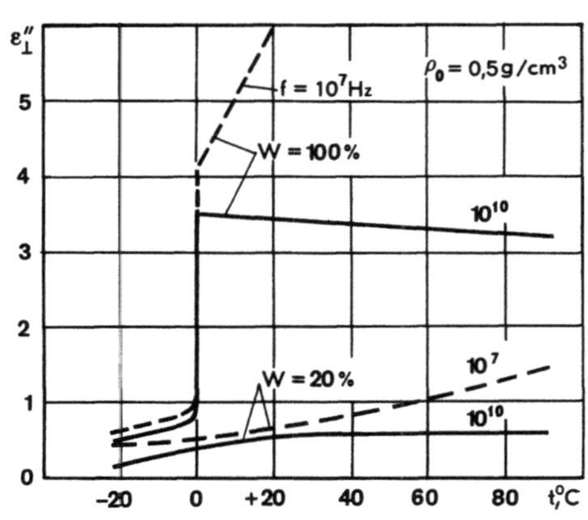

Fig. 5.21. Dependence of the loss factor $\varepsilon''_{\perp}$ on temperature; f frequency; ρ_0 density of oven-dry wood; W moisture content

5.4 Mechanism of the Influence of Moisture Content on the Dielectric Properties of Wood

In the frequency range up to 10^{10} Hz, change of the dielectric properties of dry and moist wood is due to ionic conductivity and wood dipoles (dipole-radical), interfacial and electrolytic polarization. At various frequencies, depending on the moisture content, the influence of conductivity and of different kinds of polarization on the dielectric properties changes. At low frequencies, such factors as conductivity as well as electrolytic and interfacial kinds of polarization predominate, while at high frequencies dipole polarization prevails.

The dipole-radical polarization occurs in oven-dry wood and is associated with the displacement (rotation) of OH and CH_2OH groups, and also with ionic conductivity. The latter is caused by the presence of free ions and ions of different substances absorbed by the wood molecules. The most probable relaxation time of oven-dry wood coincides with the relaxation time of the cell wall substance and is equal to 1.59×10^{-8} s at 20 °C, which corresponds to the maximum value of the loss factor at a frequency of 10^7 Hz. The ability of the polar groups to move better along the fibers in comparison with transverse movements results in higher values of the dielectric properties at an orientation of vector $\bar{E}$ along the longitudinal axis of the cells.

The influence of the moisture content on the dielectric properties of wood is the decisive factor in the entire frequency under consideration here. It is useful to analyze this influence in the following four regions of moisture content, as outlined in Chap. 2: from 0 to 5% (monomolecular moisture), from 5 to 18%–23% (adsorption moisture), from 18–23% to the fiber saturation point (capillary-condensed moisture), and above 30% (free moisture).

As the character of the interaction of different kinds of moisture with wood is different, every type of bond has its own kind of influence on the electric properties of moist wood. As a rule, at the boundary lines of the above moisture regions, significant changes of the dielectric propecties of wood occur.

The concept of the mechanism of the cellulose oscillations that take place under the influence of an electromagnetic field is based on the possibility of rotation of the three hydroxyl groups, while the glucosidic oxygen between the glucose residues remains immovable. In the oven-dry state, the macromolecules of cellulose in wood are mutually bound by secondary valence forces (Fig. 5.22a) that prevent the dipoles of the molecules from displacing under the electromagnetic field action. The process of humidification leads to the penetration of molecules of water between the molecules of cellulose and to the weakening of the transverse bonds (Fig. 5.22b), which results in an increased mobility of the dipoles.

With an increase of the wood moisture content to 1%, every 10th glucose ring or every 30th OH group in the cellulose molecule acquires 1 molecule of water. At a moisture content equal to 3%, every 10th OH group will be saturated. The influence upon the wood dielectric properties of the monomolecular water present in the wood in quantities of up to 5% (at 20 °C) is

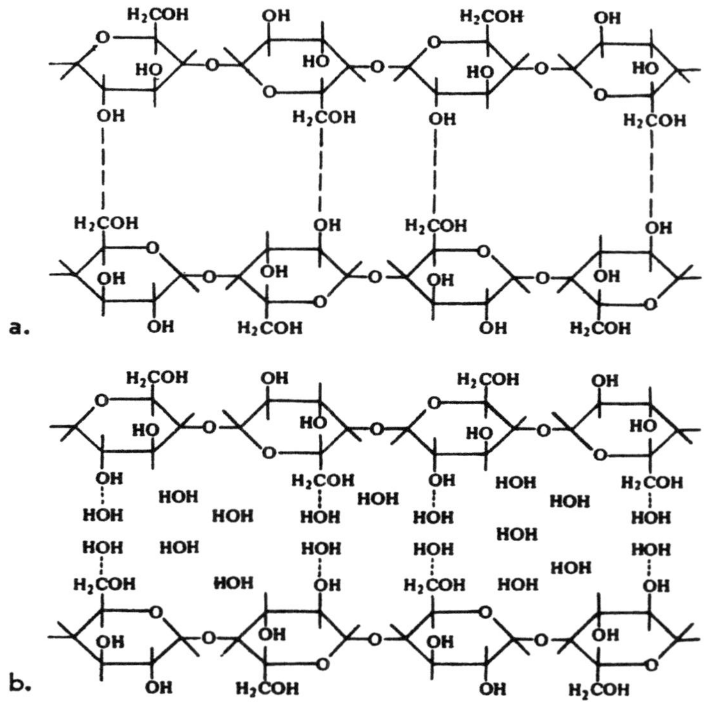

Fig. 5.22a, b. Influence of moisture content on molecular bonds of cellulose. **a** Cellulose macromolecules are connected by secondary valence forces; **b** bonds between the chains are broken by molecules of water

insignificant. This fact can be explained by the action of two factors: the high strength of the hydrogen bonds between wood and water and the relatively small quantity of moisture in the wood. However, an increase of temperature increases the dielectric properties considerably at frequencies of 20–10^3 Hz, which is explained by a corresponding increase in the conductivity of the water.

When the moisture content is increased from 5 to 18–23%, i.e., in the adsorption region, the value of the DC increases considerably (Figs. 5.1, 5.2), and an abrupt increase of the $tg\delta$ takes place (Figs. 5.12, 5.14). This increase occurs over the entire frequency range, especially at low frequencies. At these frequencies, this rise is associated with a sharp increase of conductivity, while at high frequencies, it is caused by the increase of the number of dipoles and by an enhanced dipole and interfacial polarization.

With increasing adsorption moisture in the wood, the maximum value of the loss factor, at an orientation of vector $\bar{E}$ in the radial direction and at $t = 20\,°C$, shifts from a frequency of 2.5×10^7 Hz at a moisture content $W = 3\%$ to a frequency of 8×10^9 Hz at $W = 20\%$. Table 5.22 presents values of the relaxation time τ_1 and frequency f_m that correspond to the maximum value of

Table 5.22. Relaxation time τ_1 and frequency f_m corresponding to a maximum value of the loss factor $\varepsilon''_\perp$ of spruce wood (Trapp and Pungs 1956b)

Moisture content (%)	3		6.5		12.5		20	
t	f_m (Hz)	τ_1 (s)	f_m (Hz)	τ_1 (s)	f_m (Hz)	τ_1 (s)	f_m (Hz)	τ_1 (s)
20 °C	2.5×10^7	6.4×10^{-9}	2×10^8	8×10^{-10}	10^8	1.6×10^{-10}	8×10^9	2×10^{-11}
40 °C	10^8	1.6×10^{-9}	6.9×10^8	2.3×10^{-10}	5×10^9	3.2×10^{-11}	–	–
60 °C	4×10^8	4×10^{-10}	1.4×10^9	1.1×10^{-10}				
80 °C	1.4×10^9	1.1×10^{-10}	5×10^9	3.2×10^{-11}				
100 °C	3×10^9	5.3×10^{-11}						

the loss factor at various temperatures and moisture contents (Trapp and Pungs 1956b). With increasing temperature, the maximum value of the loss factor shifts to higher frequencies, while the relaxation time τ_1 decreases. The maximum of the loss factor at $t = 20$ °C shifts with an increase of moisture content to 10% to a frequency of 10^8 Hz, and at a moisture content exceeding 20%, it is displaced in the frequency region above 10^{10} Hz.

Capillary-condensed moisture increases the $tg\delta$, but its influence is less pronounced that that of the adsorption moisture (Fig. 5.15); its effect is nevertheless larger than the influence of the free moisture.

The free moisture also increases the dielectric properties of wood and in proportion to the amount present. At some frequencies (10^6 Hz), the curve is neatly linear. The loss tangent value increases in the frequency range up to 10^8 Hz and at moisture contents from 30 to 60% (Fig. 5.15). A further increase in wood humidity up to 100% practically does not affect the value of $tg\delta$. At higher frequencies, in the range from 3×10^8 Hz to 10^{10} Hz, an increase in moisture content from 30 to 100% increases or decreases the value of $tg\delta$ depending on the relation of the $tg\delta$ values of water and wood at the fiber saturation point.

With an increase in temperature, the DC of free water at frequencies up to 5×10^9 Hz is reduced from 78 at 20 °C to 50 at 95 °C, which decreases the DC value of moist wood. At frequencies of 10^{10} Hz and higher, the DC of water increases with increasing temperature, which causes an increase in the DC of the wood.

The maximum value of the loss tangent of the free water at 20 °C is at a frequency of about 6×10^{10} Hz and is equal to 1.77 (Rosenberg 1979). With such a large value of the $tg\delta$ of water, its share in the wood $tg\delta$ value at moisture contents above 30% must be considerable, and the $tg\delta$ maximum of wood is to be expected at frequencies around 6×10^{10} Hz. However, the maximum value of the loss factor at a moisture content above 30% and at a temperature of 20 °C occurs at a frequency of about 1.5×10^{10} Hz, and at $t = 40$ °C it takes place at 2.4×10^{10} Hz.

At low frequencies (up to 10^4 Hz) wood with a moisture content above 10% has extremely large DCs and loss tangents. Thus is due to two factors, namely, the high electric conductivity of moist wood and its ability to polarize under the influence of an external electric field.

The electric conductivity of wood depends on the concentration of ions and on their mobility. An increase in moisture content augments both the quantity of dissociated ions and their possibilities for displacement, which results in a rise of the wood conductivity. James (1975) associates the conductivity of moist wood with the amorphous regions of cellulose. In his opinion, the crystalline regions are not penetrated by the water molecules, and in the presence of water they remain nonconducting, whereas the amorphous regions absorb moisture and become conductive zones. Under the influence of an external electric field, ions migrate through the amorphous regions and accumulate on the surface of the crystallites.

Wood is a heterogenous material, and to simplify any explanation of its dielectric properties, it can be pictured as a substance that on the microlevel consists of cellulose layers (and of other substances forming wood) and of bound moisture layers, whereas at the macrolevel, it consists of layers of cell walls and moisture layers which partially or completely fill the cell cavities. At the same time, the elements that separate the wood moisture have a considerably lower electric conductivity than the liquid itself. In other words, the material consists of conductive layers separated with insulating layers, and thus a multitude of the very small-sized capacitors is formed. These capacitors are connected in parallel, which makes a system of high resultant capacity and high DCs.

At low frequencies, electrolytic and interfacial polarizations take place in the wood. These kinds of polarization are characterized by large values of the polarization time constant. The influence of these kinds of polarization on the dielectric properties of wood with a moisture content exceeding the fiber saturation point seems to be insignificant in comparison with the influence of the conductivity of the water component.

At frequencies above 10^4 Hz, the influence of conductivity of the dielectric properties of moist wood gradually decreases, but it increases the role of the interfacial and dipole kinds of polarization. At frequencies above 10^8 Hz, the relaxation losses exceed the losses specified by the conductivity of the direct current, and in this case the greatest influence is exerted by the dipole polarization of water. However, it is impossible to give a quantitative estimation of the influence of different kinds of polarization and of the action of the conductivity of the direct current on the values of the dielectric properties of wood.

A plotting of the correlations of moist wood $\varepsilon'' = \varphi(\varepsilon')$ at positive temperatures by using experimental data does not permit one to obtain Cole-Cole diagrams with sufficient accuracy and thus to find the analytical dependence of ε' and ε'' values of moist wood on the frequencies, as has been done for oven-dry wood. This conclusion may be explained by the variety of dielectric properties of different forms of moisture in the wood. This variety of properties is created both by the conductivity processes and by different kinds of polarization.

At negative temperatures, the dielectric properties of wood also depend strongly on the moisture content of the wood. At $t = -50\,°C$ and with a moisture content above the fiber saturation point, about 12% of the nonfreezing moisture is still present (this moisture is a a part of the bound moisture). The remaining moisture is present in the wood in the form of ice. With an increase of temperature, the quantity of defrosted, bound moisture increases, and at a temperature of about $-1\,°C$, the bound water melts completely. A further increase in temperature up to 0 °C defrosts the free moisture completely. With the change of temperature, the ratio of ice and nonfreezing moisture changes constantly.

The ice originating from free moisture does not have a monolith structure but is formed of separate crystals. On the surface of these crystals, some quantity of water is always present in the form of a thin film, as can be shown by nuclear magnetic resonance spectroscopy. The dielectric properties of these films of water differ from those of ice. The maximum dielectric loss of water present in the films lies in the frequency range of 3×10^8–3×10^9 Hz, whereas the maximum value of ice loss is at a frequency of 3×10^4 Hz.

The dielectric properties of the cell wall, ice, nonfreezing, bound moisture, and monomolecular water films are all different. The quantitative relation between the water components in wood depend not only on the temperature but also on the direction of the process as well, namely, whether freezing or melting is under consideration. Naturally, this is the reason why the coefficient of variation of the dielectric properties of wood at negative temperatures has such a high value.

At moisture contents up to 30%, an increase in temperature in the negative region causes a continuous increase of the DC. This DC value is not subject to abrupt changes when passing the 0 °C point. At moisture contents above 30%, the presence of free moisture causes an increase of the DC at 0 °C. The influence of conductivity on the dielectric properties of frozen, moist wood is several times lower than its influence at positive temperatures, and processes of polarization therefore play the main role at frequencies above 10^4 Hz.

The maximum values of the loss factors of the components of frozen wood occur in different frequency ranges. This creates possibilities for the efficient use of electromagnetic field actions for the reduction in the strength of frozen wood by selectively influencing the bound moisture as well as the intercrystalline moisture in the films.

The process of dielectric relaxation in substances consisting of dipole molecules is a combination of the rotary vibration of molecules (or groups) relative to certain temporary positions of equilibrium and subsequent quick and abrupt reorientations of the molecules (groups) which causes their transition to a new position of equilibrium. Such a transition is accompanied by a change of position of the molecules (groups) relative to the surrounding ones. The reorientation of molecules (groups) acquires the character of activated jumps over the energy barriers. This is the reason why the dielectric relaxation process is considered under the theory of the absolute reaction rate. The process of the transition of the molecule or group from one position of equilibrium to another requires a certain free energy of activation for crossing the energy barrier that separates these positions of equilibrium. The free energy of activation of the dipole relaxation is associated with the most probable time of dielectric relaxation τ_m, the value of which corresponds to the frequency f_m of the maximum value of dielectric losses.

The principal energy factor characterizing the process of dielectric relaxation in dry and moist wood is the apparent activation energy. Several investigators (Trapp and Pungs 1956b, Tsutsumi 1967, Norimoto 1976, Benkova 1981, Handa et al. 1982) have shown that the dependence of $\ln \tau$ on $1/T$ for oven-dry and moist woods has a linear character. In this case, the activation energy can be calculated by employing well-known equations presented in works by Akhadov (1972), Norimoto (1976), and Skaar (1988) on the basis of empirical angle coefficient values in the dependence of $\ln \tau$ on $1/T$.

The apparent activation energy ΔE of oven-dry wood of various tree species according to Norimoto (1976) in the temperature range from -69 to $53\,°C$ changes from 32.3 to 42.7 kJ/mol. It should be noted that with an orientation of the vector of the electric field strength along the fiber direction, the value of ΔE is 13% greater than that with a transverse orientation. The activation energy of ice in the temperature range from 0 to $-50\,°C$ amounts to 42.7–57.8 kJ/mol, and that of free water at temperatures from 0 to $100\,°C$ is 13.4–21.4 kJ/mol. The ΔE value of monolayer water that is sorbed on the wood is around 34 kJ/mol, and that of polylayer water in the temperature range from -50 to $80\,°C$ lies within 8.4–12.6 kJ/mol (Kolosovskaja et al. 1989). According to various authors, the value of the apparent activation energy of different woods with a moisture content of 6–16% and in the temperature range from -40 to $80\,°C$ lies within 34–59 kJ/mol, with an average of 55 kJ/mol, which is near to the ΔE of ice.

Chapter 6

Effect of Different Kinds of Treatment on the Dielectric Properties of Wood

In order to impart to wood special properties, it is subjected to mechanical, physical, and chemical kinds of treatment. These procedures change its macrostructure and its molecular composition and affect the natural bonds between its components which all results in changes of the conductivity and polarizability of the wood. If the wood is compacted, its density changes depending on the degree of pressure. Wood is also subjected to radioactive irradiation and to different kinds of chemical treatment for the purpose of increasing its structural strength or resistance to burning, changing its electrical properties, improving its life span, etc. These processes simultaneously change the dielectric properties of the wood.

6.1 Dielectric Properties of Compressed Wood

Under the action of external pressure exceeding the compressive strength of the wood, the volume of the pores diminishes, and the wood density increases. As a rule, the compression is carried out in a direction perpendicular to the orientation of the fibers, i.e., in the directions *R* and *T*, or at an angle to them. This process distorts the structure of the wood. The extent of change of the wood density value and destruction of the cell wall depend on the rate of compression.

Compression in the radial direction changes the configuration of the cells in the early part of the annual rings of coniferous and ring-porous woods. The latewood elements change only when highly compressed. However, full collapse of the tracheid walls of coniferous wood cannot be attained even at extremely high pressures. The rays in the early part of the annual rings in coniferous and ring-porous woods acquire a zigzag form, but in the late part, their form does not change.

Compressing in the tangential direction of coniferous and ring-porous woods results in a zigzag form of annual layers. Compressing the diffuse-porous species of wood changes their microscopic constitution in the whole cross-section of the annual layer. On radial compression, the vessels flatten out and stretch into zigzag voids, and on tangential compression, the vessels acquire the form of elongated narrow slits. The degree of compression μ_1 in the direction of force application can be determined with the following equation, in percent

$$\mu_1 = (h_0 - h) \times 100/h_0 \,, \tag{6.1}$$

where

h_0 = original dimensions of the sample
h = dimensions of the sample after compression.

A compression of wood to 60% of its original dimensions does not cause structural changes of the cellulose fibers. As a result of such a compression, only a reorientation of the fibers in the cell walls takes place. Such changes of structure affect the dielectric properties of compressed wood.

Experiments carried out by a number of authors (Efremov and Zhilenkov 1954, Miagkov 1954, von Hippel 1954, Rafalski 1966, Boievskaja et al. 1986) showed that with increased compression of oven-dry wood, its dielectric properties increase over the entire range of frequencies from 20 to 10^{11} Hz. This can be attributed largely to an increase of the density of the wood.

The frequency responses of the dielectric constant (DC) of wood, the loss tangent, and the loss factor values with vector $\bar{E}$ oriented perpendicular to the fiber direction of oven-dry Douglas-fir wood with a density of 1.4 g/cm^3, at $t = 25\,°C$, plotted according to data obtained by von Hippel (1954) are presented in Fig. 6.1. With increasing frequency, the diminishing of $\varepsilon'_\perp$ from 7.3 at a frequency of 10^2 Hz to 3.2 at $f = 2.5 \times 10^{10}$ Hz takes place. The character of the decline of the DC of compressed wood is similar to that for the cell wall substance, but the absolute value of $\varepsilon'_\perp$ of compressed wood is higher, even though its density is lower.

The maximum value of the loss tangent of compressed wood with a density of 1.4 g/cm^3 is at a frequency of about 10^8 Hz, its value being 0.085 (whereas the cell wall substance $tg\delta_\perp = 0.068$). The point of loss factor maximum is more definitely expressed, and its value is 0.42. The frequency at which this point is achieved is equal to 10^7 Hz, and this is the same frequency at which the maximum value of $\varepsilon''_\perp$ of the cell wall substance is found (its maximum value is equal to 0.34). The maximum of the loss factor in comparison with that of the loss tangent is shifted in the direction of lower frequencies.

The higher values of the dielectric properties of oven-dry, compressed wood with a density of 1.4 g/cm^3 in comparison with those of the cell wall substance (whose density is equal to 1.53 g/cm^3) can be explained by the changes of wood

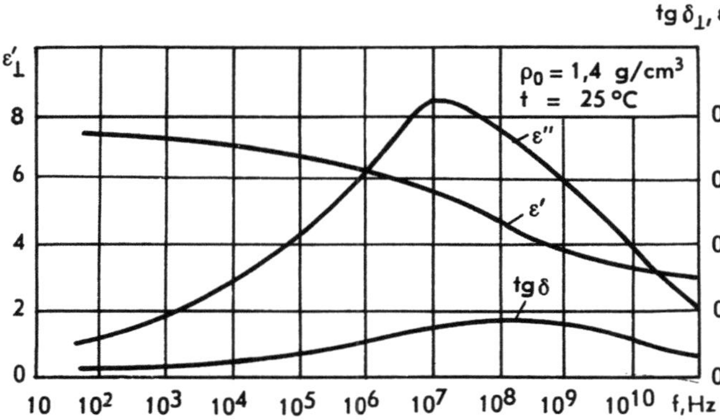

Fig. 6.1. Frequency response of dielectric parameters $\varepsilon'_\perp$, $tg\delta_\perp$, and $\varepsilon''_\perp$ of oven-dry compressed softwood (von Hippel 1954)

structure and the chemical alterations of the substances forming the wood that are caused by the temperature of compression. These changes increase the values of the wood dielectric properties.

At low frequencies and increased degree of compression, the value of $\varepsilon'_\perp$ of oven-dry wood increases almost linearly. According to Miagkov (1954), the correlation between the values of ρ_0 and DC is linear. This correlation, at a frequency of 50 Hz, with $t = 20\,°C$, and with values of ρ_0 from 0.5 to 1.1 g/cm^3, is expressed by the following empirical equations:

For birch wood $\varepsilon'_\perp = 2.45\rho_0 + 1.17$
For alder wood $\varepsilon'_\perp = 2.02\rho_0 + 1.33$
For aspen wood $\varepsilon'_\perp = 1.54\rho_0 + 1.78$
For beech wood $\varepsilon'_\perp = 2.7\rho_0 + 0.88$
For pine wood $\varepsilon'_\perp = 3.2\rho_0 + 0.95.$

In the high frequency range the dependence of ε', $tg\delta$, and ε'' of wood compressed in the transverse direction upon the density is nearly linear, whereas their dependence on the compression rate is curvilinear (Rafalski 1966). In the frequency range from 0.2 to 25 MHz, the DC value of oven-dry red beech wood decreases continuously with increasing frequency (Fig. 6.2) at any compression rate, whereas the loss tangent in the frequency range from 0.2 to 4 MHz sharply increases. Above this frequency and up to 25 MHz, its values increase almost linearly.

The values of ε' of compressed wood at a compression rate equal to 37.5% in the radial direction are on average 5% higher than those when the wood is

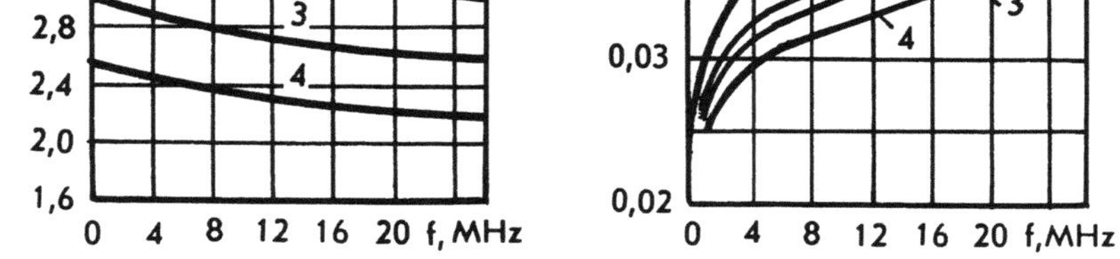

Fig. 6.2a, b. Dependence of ε'_R and $tg\,\delta_T$ values of oven-dry red beech at various degrees of compression over a frequency range of 0.2–25 MHz, at $t = 20\,°C$ (Rafalski 1966). **a** Vector $\overline{E}$ oriented radially, **b** vector $\overline{E}$ oriented tangentially. *1*, *2*, *3*, and *4* compression degrees representing 50% ($\rho_0 = 1.4$ g/cm^3), 37.5% ($\rho_0 = 1.23$ g/cm^3), 25% ($\rho_0 = 1.05$ g/cm^3), 0% ($\rho_0 = 0.7$ g/cm^3), respectively

compressed in the tangential direction. With an increase in the compression rate up to 50%, the difference between the two values virtually disappears, which can be explained by a reduction in the influence of the rays. In the frequency range of 5–25 MHz, the loss tangent value in the radial direction is on average 8% higher than that in the tangential direction, and its changes with the variation in frequency and compression rate are insignificant.

Rafalski (1966) has suggested that the average values of $\varepsilon'_\perp$ and $tg\delta_\perp$ at $f = 10^7$ Hz of compressed, oven-dried red beech wood can be computed using the following empirical equations:

$$\varepsilon'_\perp = 2.16\rho_0 + 0.77 \tag{6.2}$$

$$tg\delta_\perp \times 10^4 = 78.32\rho_0 + 261 \,, \tag{6.3}$$

where

ρ_0 = density value of oven-dry compressed wood (g/cm^3).

The dielectric characteristics of oven-dry compressed wood of some tree species at high frequencies are given in Table 6.1. This table is based on data reported by Miagkov (1954) and Boievskaja et al. (1986).

Comparison of the values of the DC $\varepsilon'_\perp$ of oven-dry, compressed wood obtained by various investigators at frequencies from 50 to 10^{10} Hz and with wood density values from 0.5 to 1.4 g/cm^3 with the data in Appendix 1 relating to unchanged wood shows that, at a corresponding density, they differ within the limits of -20% to $+25\%$. Accordingly, for estimation of the DC of oven-dry compressed wood with the vector $\bar{E}$ oriented perpendicular to the fiber direction, the data in Appendix 1 can be used in cases when the density of the unchanged wood is equal to the density of the compressed wood. The values of $tg\delta_\perp$ of oven-dry compressed wood in the frequency range from 50 to 10^{10} Hz differ from those of unchanged wood given in Appendix 1 within $\pm 35\%$, provided the density values are equal to each other.

Table 6.1. Dielectric parameters $\varepsilon'_\perp$ and $tg\delta_\perp$ of compressed wood of certain species at $t = 20\,°C$ and at frequencies of 10^6 and 10^7 Hz (Miagkov 1954, Boievskaja et al. 1986)

Species	Density (%)	W (%)	$f = 10^6$ Hz		$f = 10^7$ Hz	
			$\varepsilon'_\perp$	$tg\delta_\perp$	$\varepsilon'_\perp$	$tg\delta_\perp$
Birch	0.73	0	2.85	0.034	2.85	0.055
	0.97	0	3.0	0.032	3.3	0.056
	1.12	0	3.3	0.043	3.4	0.057
	1.40	0	4.1	0.044	3.96	0.051
	1.40	6.5	5.34	0.055	4.87	0.063
Aspen	0.93	0	3.33	0.040	3.25	0.058
Alder	0.88	0	2.82	0.032	2.75	0.053
Lime	0.83	0	2.76	0.037	2.70	0.050
	0.89	0	2.95	0.036	2.95	0.056
Pine	0.94	0	3.3	0.030	3.0	0.073

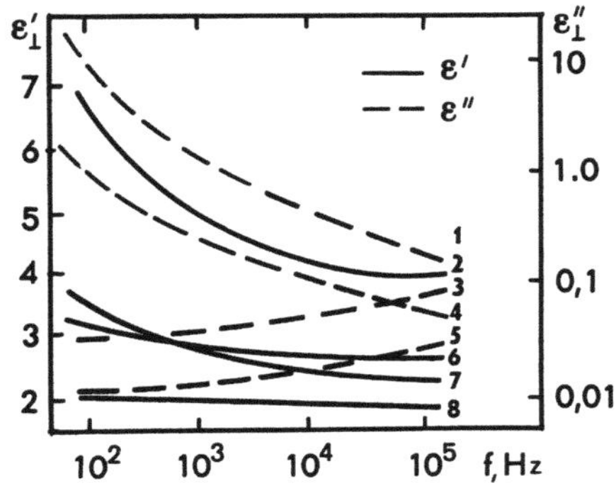

Fig. 6.3. Frequency response of the dielectric constant $\varepsilon'_\perp$ and loss factor $\varepsilon''_\perp$ of compressed red pine wood at low frequencies (Venkateswaran): *1, 2* $W = 12.5\%$, density $\rho = 0.8$ g/cm^3; *3, 6* $W = 0\%$, $\rho_0 = 0.78$ g/cm^3; *4, 7* $W = 12.5\%$, $\rho = 0.41$ g/cm^3; *5, 8* $W = 0\%$, $\rho_0 = 0.4$ g/cm^3

Figure 6.3 shows the influence of the moisture content on the dielectric properties of compressed wood at low frequencies. An increase of moisture content of compressed red pine wood ($\rho_0 = 0.8$ g/cm^3) from 0 to 12.5% increases the $\varepsilon'_\perp$ value at $f = 10^2$ Hz 2.1 times, and at $f = 10^5$ Hz 1.5 times. The loss factor of compressed wood with a moisture content $W = 12.5\%$ diminishes by almost two orders of magnitude with an increase of frequency from 10^2 to 10^5 Hz.

Venkateswaran (1973) notes that at a frequency of 10^3 Hz compressed wood with a moisture content 0–12.5% has a very close, linear correlation between the density value (within 0.4–1.1 g/cm^3) and DC in the transverse direction. The correlation coefficient for compressed wood is equal to 0.991–0.997, values that are much higher than those (for unchanged wood) given by Skaar (1948) and Hearmon and Burcham (1954), whose values are 0.911–0.933 and 0.723–0.931, respectively.

At frequencies of 10^5–10^7 Hz, an increase of moisture content from 0 to 6.5% increases the DC value of compressed birch wood ($\rho_0 = 1.4$ g/cm^3, $t = 20\,°$C) $\varepsilon'_\perp$ 23–32% and $\varepsilon'_\parallel$ 16–37%. The value of $tg\delta_\perp$ increases 24–34%, and that of $tg\delta_\parallel$ 3–8%. At the same frequencies and at a moisture content of 0–6.5%, the value of DC in the longitudinal direction is higher than that in the transverse direction by 15–20%, while the loss tangent value is 1.4–1.7 times higher.

At a compression of 30%, at a temperature of $t = 20\,°$C, and at frequencies of 10^6–10^7 Hz, an increase in moisture content from 3 to 63% increases the DC value in the transverse direction from 2.1–2.4 to 4.1–4.5 for pine wood (Efremov and Zhilenkov 1954). An increase of the compression of pine wood from 20 to 40%, at W equal to 20%, and at the same frequencies, increases the values of $\varepsilon'_\perp$ 1.7 times, and those of $tg\delta_\perp$ 1.4–2.7 times.

The dependence of the dielectric characteristics of compressed wood on the wood density and moisture content and on the frequency is similar to that of unchanged wood. However, the absolute values of the properties of compressed wood are higher than those of unchanged wood. The influence of temperature on the dielectric properties of compressed wood has not been studied.

6.2 Dielectric Characteristics of Modified Wood

Wood is often treated with various chemical substances in order to change its mechanical and physical properties. Treatment of wood with synthetic resins diminishes its swelling 2–4 times, increases its strength 2–5 times, and increases its hardness up to 78 times. In addition, this treatment stabilizes the electric properties of wood and extends its service life. Intensification of the processes of wood modification and an increase of quality of wood-based materials can be achieved by the use of electromagnetic fields of high and super high frequencies. These methods make it possible to carry out the process of polymerization in minutes instead of hours, as required when convection methods are used.

6.2.1 Wood Treated with Ammonia

Modification of wood with ammonia is intended to enhance the ability of wood to be deformed, as well as to raise its density in the process of drying and to increase the penetration of paint or stain. The duration of the process of wood modification with ammonia to a great extent depends on the rate of heating of the material. The use of dielectric methods of heating makes it possible to intensify the process. In order to carry out this process properly, it is necessary to know the dielectric properties of wood containing ammonia.

Klotinj and Darsinsh (1983) studied the influence of the ammonia content in the wood on its dielectric properties. They measured the dielectric properties of gray alder wood containing from 0.2 to 7.5% of ammonia in relation to the oven-dry wood mass. The experiments were carried out at 20 °C in the frequency range from 0.8 to 20 MHz, with wood containing 15% moisture and with vector $\bar{E}$ oriented in the radial direction.

It was established that an increase of the ammonia content causes a significant increase in the dielectric properties of wood (Fig. 6.4). At a frequency of 0.8 MHz, an increase in ammonia content from 0.2 to 7.5% increased the DC value from 5.3 to 29, and at a frequency of 20 MHz, this increase was from 4.6 to 6.7. Simultaneously, the $tg\delta_R$ value, at a frequency of 0.8 MHz increased from 0.14 to 1.35, and at frequency 20 MHz from 0.07 to 0.29. When the frequency was increased from 0.8 to 20 MHz, the values of ε'_R and $tg\delta_R$ of wood impregnated with ammonia decreased considerably.

By the end of the process of removing the ammonia from the wood, the residual content of ammonia in the wood is equal to approximately 0.2%, and the dielectric constant of such wood does not differ from that of the original wood. At the same time, the loss tangent value increases about 1.4–1.9 times.

The increase of the dielectric properties of wood impregnated with ammonia is due to the fact that the gaseous ammonia dissolves in the water in the wood, forming ammonium hydroxide whose dissociation increases both the ionic conductivity of the wood and the structural polarization.

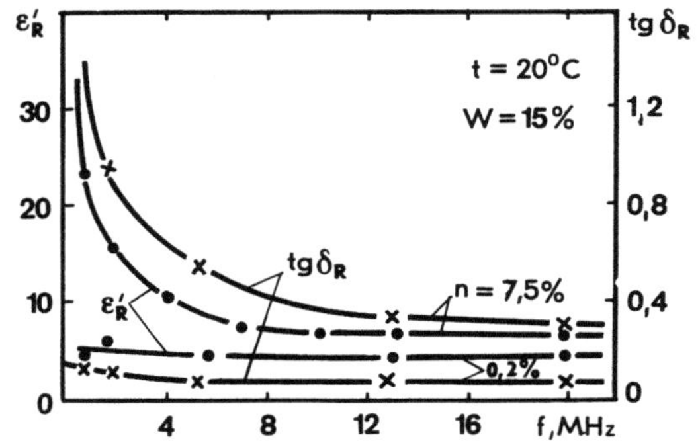

Fig. 6.4. Frequency response of ε'_R and $tg\delta_R$ of alder wood impregnated with ammonia (Klotinj and Darsinsh 1983): f frequency; n quantity of ammonia in the wood (percentage of the oven-dry wood mass); W moisture content

By treating wood with ammonia and compressing it with subsequent drying, plasticized wood is produced. This material is called lignamon. It has higher dielectric properties than the original wood both in the longitudinal and in the transverse directions. Table 6.2 shows values of ε' and $tg\delta$ of lignamon from birch of different densities and measured at 50 Hz.

The DC and loss tangent values of plasticized wood at a frequency of 50 Hz increase with increasing wood density. When the moisture content is increased from zero to the fiber saturation point, the DC value of plasticized wood increases 1.5–2 times along the fibers and 1.4–1.7 times in the transverse direction. The value of $tg\delta$ increases 18–75%.

At high frequencies, the DC value of plasticized wood with the orientation of vector $\bar{E}$ transverse to the fiber direction decreases with increasing frequency from 10^5 to 2.5×10^6 Hz and at a moisture content of 0% for wood with

Table 6.2. Dielectric constant and loss tangent values of birch wood plasticized with ammonia and compressed, at f = 50 Hz and t = 20 °C (Postnikov 1972)

Moisture content (%)	Density (g/cm³)	$\varepsilon'_\perp$	$tg\delta_\perp$	$\varepsilon'_\parallel$	$tg\delta_\parallel$
0	0.75	2.6	0.07	3.9	0.08
0	1.02	4.9	0.08	4.9	0.09
0	1.2	6.1	0.10	5.8	0.11
0	1.31	6.2	0.10	6.2	0.12
27.2	0.75	7.1	0.12	7.9	0.11
26.0	1.02	8.0	0.14	8.2	0.12
23.5	1.2	8.5	0.13	8.9	0.13
21.3	1.31	9.1	0.13	9.1	0.13

a density of 0.75 g/cm^3 from 3.2 to 2.0, and for wood with a density of 1.2 from 6.5 to 2.9. At a moisture content of 23–27% and a density of 1.2 g/cm^3, the DC value decreases from 7 to 3.7. The loss tangent value at a frequency of 10^6 Hz and at zero moisture content for wood with densities of 0.75–1.02 g/cm^3 is 0.03–0.05, while at a moisture content of 23–27%, the value of $tg\delta_\perp$ lies within 0.06–0.09 (Postnikov 1972). With vector $\bar{E}$ oriented parallel to the fiber direction, the $tg\delta$ is 19–25% higher than that in the transverse direction.

6.2.2 Dielectric Properties of Wood Impregnated with Various Substances

Modification of wood by saturating it with various impregnating substances (monomers, oligomers, resins) with their subsequent hardening is intended for improvement of the performance characteristics of untreated wood. The change of wood properties is caused by the reduction in the volume of the cell cavities by their partial filling with the polymers and by formation of polymers in the cell walls of wood. The location of polymers in wood depends on the nature of the filler and on the impregnation method. The change of the wood dielectric properties as a result of such modifications is determined by the structure and properties of the components and by the quantity of the filler and its interaction with the wood.

The amount of the filler in the wood is generally characterized by a quantity ratio (the degree of impregnation)

$$s = m_f/m_w \,, \tag{6.4}$$

where

m_f = mass of the filler
m_w = mass of the wood.

The ratio is often expressed in percent.

Glukhov et al. (1973) recommended that the DC value ε'_n of wood modified with a filler be calculated using the following equation

$$\varepsilon'_n = \varepsilon'_1 + (\rho_1 s/\rho_f)\varepsilon'_f - (\rho_1 s/\rho_f)\varepsilon'_b \,, \tag{6.5}$$

where

ε'_1, ε'_f, and ε'_b = dielectric constant values of wood, filler, and air
ρ_1 and ρ_f = density values of wood and filler
s = mass of filler related to the initial mass of wood.

The loss tangent value $tg\delta_n$ of the modified wood can be computed with the formula (similar to Eq. 2.5)

$$tg\delta_n = tg\delta_1 + V_n tg\delta_f \,, \tag{6.6}$$

where

V_n = relative volume of filler in the wood
$tg\delta_1$, $tg\delta_f$ = loss tangent values of wood and filler.

Impregnation of wood is carried out with materials (fillers) such as acrylonitrile, vinyl acetate, vinylidene chloride, methyl methacrylate, polyester resins (nonsaturated), styrene, urea-formaldehyde resins, glyptal resins, phenolic alcohols, furan resins, epoxide resins, and others. Mixtures of different substances are also used. Some of the above substances penetrate easily into the cell walls, namely phenolic alcohols and furan resins. By using, for example, special methods of impregnation, to these substances can be added acrylonitrile, methyl methacrylate, and styrene. The remaining substances penetrate only the wood cell cavities. Dielectric characteristics of several kinds of fillers for wood are given in Table 6.3.

Modification of wood with polyester resin PN-1 (33% styrene solution of polycondensation products of diethylene glycol maleic anhydride and phthalaldehyde), whose DC is higher than that of oven-dry wood, results in an increase of the dielectric properties. The values of DC and $tg\delta$ of birch wood (at high frequencies) modified with PN-1 resin are given in Table 6.4, which was compiled from data reported by Olekhnovich et al. (1974). At frequencies of 10^5–10^7 Hz, the DC value in the longitudinal direction is 10–17% higher than in the transverse direction, while the $tg\delta_{||}$ value is higher by 50–70%.

The values of the dielectric properties of wood modified with PN-1 resin are intermediate between those of the resin and the wood. In the process of

Table 6.3. Dielectric properties of some wood fillers at frequency 10^6 Hz, $t = 20\,°C$ (Rotsens et al. 1983)

Filler	ε'	$tg\delta$
Polyacrylonitrile	4.2	0.03
Polyvinyl acetate	3.1	0.0045
Polyvinyl chloride	3.2–3.3	0.018–0.019
Polymethyl methacrylate	2.6–3.8	0.02–0.03
Polystyrene	2.5–2.6	0.0002–0.0004
Polyethylene	2.2–2.3	0.0002
Polyester resins	2.8–5	0.010–0.045
Furan resins (hardened)	3.5–6	0.008–0.0185
Epoxide resins (hardened)	3.5–6	0.01–0.03

Table 6.4. Dielectric properties of birch wood modified with resin, PN-1, at $t = 20\,°C$, $W = 5$–6%, $\rho = 1.1$ g/cm^3 (Olekhnovich et al. 1974)

	Frequency (Hz)			
	10^5	10^6	10^7	2.4×10^9
$\varepsilon'_{\perp}$	4.4–4.6	4.1–4.3	4–4.1	2.43
$tg\delta_{\perp}$	0.023–0.026	0.032–0.035	0.046–0.049	0.038
$\varepsilon'_{\|\|}$	5–5.4	4.7–5.0	4.4–4.7	2.9
$tg\delta_{\|\|}$	0.039–0.043	0.055–0.060	0.068–0.074	0.057

hardening of the resin, the dielectric properties of the material change. These changes can be revealed by a measuring instrument and may be used for inspection of the hardening process of the resin.

Phenol-formaldehyde resins penetrate into the cell walls and change their physical and mechanical properties. They possess a pronounced polarity which is caused by the presence of OH groups in the molecules of this polymer. At a temperature of 20 °C and in the frequency range from 10^2 to 10^{10} Hz, the DC value of the resin is 4.1–4.8, and that of the loss tangent is 0.010–0.028. Table 6.5 presents the dielectric properties of birch wood modified with phenol-formaldehyde resin with a resin content of 20–65% relative to the oven-dry wood. With an increase of the resin content from 20 to 32%, the value of the DC increases 2–7%, and the value of $tg\delta$ 2–15%. An increase of resin content from 20 to 62–65% increases the DC value 13–27%, and the value of $tg\delta$ 4–31%.

With an increase in frequency from 10^3 to 2.3×10^7 Hz, the DC value of wood modified with phenol-formaldehyde resin decreases continuously, whereas the value of $tg\delta$ increases linearly. The relation $k_1 = \varepsilon'_{\parallel}/\varepsilon'_{\perp}$ changes within 1.36–1.21, and the relation $k_2 = tg\delta_{\parallel}/tg\delta_{\perp}$ within 1.80–1.33. With increasing proportion of resin in the wood, the difference between the dielectric properties in the longitudinal and in the transverse directions diminishes.

The dielectric properties of modified wood are influenced not only by the chemical composition of the polymer but also by its structure. Impregnation of wood with low-molecular polyethylene to a volume of polyethylene equal to 68–70% relative to the mass of the samples in the oven-dry state increases the DC value of wood. The DC value of polyethylene depends upon its density. With a change in the polyethylene density, its DC value changes from 2.2 to 2.45, and practically does not change with a change in frequency from 50 to 10^{10} Hz. The value of the DC of wood impregnated with polyethylene is higher than that of wood and polyethylene alone. This can be partly explained by the fact that the polyethylene that fills the wood capillaries displaces their air. The DC of air is lower than that of polyethylene, and thus the DC of the composite increases. In addition, the layers of cell wall substance alternate with the layers of polyethylene and form many capacitors that can be considered as connected in parallel. Such a system of capacitors increases the total capacitance, which results in a considerable increase of the dielectric characteristics of the composite. The data in Table 6.6 (Olekhnovich et al. 1975) show the dependence of the dielectric properties of wood on the frequency and electric field orientation relative to the fiber direction.

The loss tangent value of low molecular weight polyethylene at a frequency of 50 Hz is 3–4 times higher than that of high molecular weight polyethylene and is equal to 0.0012. With increasing frequency, its value increases, and at $f = 10^6$ Hz it reaches a value of 0.0033. The value of the loss tangent of modified, oven-dry wood at a frequency of 50 Hz, in comparison with that of unchanged wood, decreases with the orientation of field vector $\bar{E}$ transverse to the fiber direction and increases with a longitudinal orientation. At frequencies of 10^6–10^7 Hz, the value of $tg\delta$ increases and becomes as high as 0.053–0.058. The

Table 6.5. Dielectric parameters of birch wood modified with phenol-formaldehyde resin at $t = 20\,°C$ (Olekhnovich et al. 1976)

Resin content (%)	W (%)	Frequency (Hz) 10^3				2.5×10^6				2.3×10^7				2.4×10^9			
		$\varepsilon'_\perp$	$tg\delta_\perp$	$\varepsilon'_\parallel$	$tg\delta_\parallel$	$\varepsilon'_\perp$	$tg\delta_\perp$	$\varepsilon'_\parallel$	$tg\delta_\parallel$	$\varepsilon'_\perp$	$tg\delta_\perp$	$\varepsilon'_\parallel$	$tg\delta_\parallel$	$\varepsilon'_\perp$	$tg\delta_\perp$	$\varepsilon'_\parallel$	$tg\delta_\parallel$
0	0	2.37	0.009	2.90	0.015	2.08	0.027	2.56	0.049	1.88	0.036	2.35	0.054				
20	5–6	2.62	0.020	3.57	0.036	2.28	0.029	3.0	0.050	2.21	0.038	2.83	0.056				
32	5–6	2.75	0.023	3.64	0.038	2.37	0.033	3.22	0.051	2.32	0.042	3.03	0.058				
62–65	5–6	–	–	–	–	2.82	0.038	3.40	0.052	2.80	0.046	3.38	0.061				
12.5	5–6													2.5	0.047	2.9	0.064

Table 6.6. Dielectric parameters of birch wood modified with low-molecular polyethylene with a density of 0.86 g/cm^3 (oven-dry birch wood density equal to 0.605 g/cm^3), degree of impregnation by mass 68–70%, $t = 20\,°C$ (Olekhnovich et al. 1975)

Wood	Moisture content (%)	Frequency (Hz) 50				10^5				10^6				10^7			
		$\varepsilon'_\perp$	$tg\delta_\perp$	$\varepsilon'_\parallel$	$tg\delta_\parallel$	$\varepsilon'_\perp$	$tg\delta_\perp$	$\varepsilon'_\parallel$	$tg\delta_\parallel$	$\varepsilon'_\perp$	$tg\delta_\perp$	$\varepsilon'_\parallel$	$tg\delta_\parallel$	$\varepsilon'_\perp$	$tg\delta_\perp$	$\varepsilon'_\parallel$	$tg\delta_\parallel$
Original	0	2.22	0.0084	2.96	0.0115	2.1	0.02	2.56	0.029	2.1	0.0258	2.52	0.0467	2.0	0.026	2.47	0.048
Modified	0	2.92	0.0064	4.28	0.0172	2.94	0.02	3.82	0.038	2.88	0.0273	3.6	0.0526	2.76	0.03	3.4	0.058
Original	7	3.0	0.0331	6.24	0.263	3.0	0.02	4.0	0.04	3.0	0.0350	3.85	0.0595	2.9	0.06	3.7	0.089
Modified	3.5	–	–	5.54	0.112	3.6	0.02	5.1	0.04	3.5	0.0335	3.9	0.05	3.46	0.05	4.6	0.075

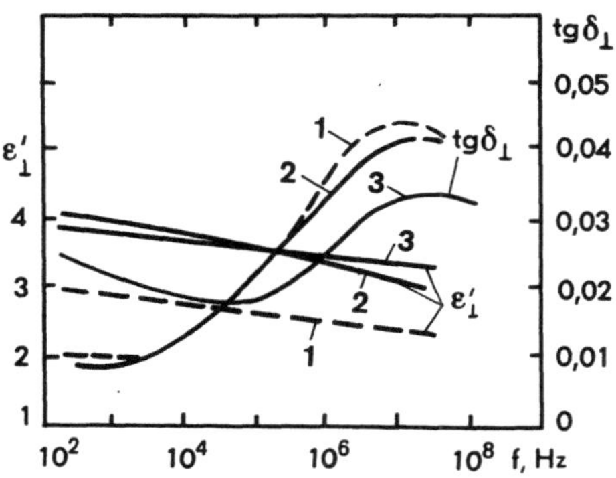

Fig. 6.5. Frequency response of $\varepsilon'_\perp$ and $tg\delta_\perp$ values of impregnated birch wood at temperature $t = 25\,°C$ and moisture content $W = 0\%$ (Glukhov et al. 1973): *1* unchanged wood; *2* wood impregnated with styrene ($\rho = 0.86$ g/cm³, $s = 54\%$); *3* wood impregnated with methyl methacrylate ($\rho = 1.1$ g/cm³, $s = 65\%$)

loss tangent value of unmodified, moist wood is higher than that of modified wood (Table 6.6). This phenomenon can be explained by the diminishing quantity of moisture in the wood as a result of a decrease in its hygroscopicity. The modification of wood with low-molecular polyethylene assures the stability of the dielectric properties of the material with changes in the environment.

The impregnation of wood with styrene, methyl methacrylate, polymethyl methacrylate, and acrylonitrile increases the DC value with increasing degree of impregnation (Fig. 6.5). With increasing degree of impregnation of birch wood with methyl methacrylate from 20% to 65%, at a temperature of 25 °C and at zero moisture content, the value of the DC (measured at $f = 10^3$ Hz) increases from 3.2 to 3.82. The DC value decreases with increasing frequency.

The maximum value of $tg\delta$ of wood impregnated with styrene ($s = 54\%$) occurs at the same frequency as that of untreated wood (Fig. 6.5), whereas for wood impregnated with methyl methacrylate it is shifted to higher frequencies; in addition, the absolute value of $tg\delta$ in this case is much lower. Treating the wood with ethanol as a swelling agent simultaneously with styrene or methyl methacrylate decreases the maximum value of $tg\delta$, which may be explained by an increase in the relaxation time range of the wood dipoles as a result of introducing the monomer into the cell walls.

Measurements of the $tg\delta$ of wood impregnated with methyl methacrylate with vector $\bar{E}$ oriented along the fibers showed that the character of dependence $tg\delta = \phi(f)$ is similar to that for the transverse orientation, but the absolute values of $tg\delta_{||}$ are higher (Glukhov et al. 1973).

The modification of wood with polymers increases the activation energy of dipole polarization, which is 32–43 kJ/mol for untreated wood, to 59 kJ/mol at a 55% degree of impregnation with styrene and up to 67 kJ/mol at a 65% degree of impregnation with polymethyl methacrylate.

6.3 Influence of Gamma Radiation on the Dielectric Properties of Wood

The action of radioactive irradiation is characterized by the energy that is released in the substance subjected to radiation. This applies to all kinds of ionizing radiation. The relation of the absorbed energy of the ionizing radiation to the mass of the substance subjected to radiation is called the absorbed dose. The absorbed dose is measured in J/kg. This unit is called a gray (Gy).

The irradiation of wood with different doses of gamma-radiation results in the destruction of all wood components on the macromolecular level. An investigation of the influence of different doses of irradiation on pine wood, which was carried out by Freidin (1961), showed that lignin is a more radioresistant component than cellulose. With increasing dose, the content of lignin diminished from 25.6% before the sample was subjected to irradiation to 18% after 5×10^6 Gy, i.e., about 26%. At the same time, the content of cellulose decreased from 49.7 to 25%, i.e., by 50%. with an absorbed dose of 5×10^5 Gy, the content of cellulose decreased by 10%. With an increase of the absorbed dose, the cellulose macromolecules becomes shorter. The major influence on the dielectric properties of wood as a result of gamma-radiation are the changes that occur with cellulose and the hemicelluloses.

According to investigations of Glukhov et al. (1973) and Norimoto and Yamada (1972, 1975) in the frequency range from 10^2 to 10^8 Hz at an absorbed dose equal to 10^5 Gy, the dielectric characteristics of wood and cellulose do not suffer any significant changes. A dose equal to 3×10^4 Gy increases the dielectric properties of cellulose slightly and causes minor changes in its degree of crystallinity. The values of ε' and ε'' increase notably with a radiation dose exceeding 29×10^4 Gy and reach their peak value at about 10^6 Gy after which these values decrease sharply when the irradiation dose is increased further. The dependence $\varepsilon'' = \phi(f)$, at a dose equal to 3.2×10^6 Gy, undergoes minor changes with an alteration of frequency from 4×10^2 to 10^6 Hz. It is assumed that the number of CH_2OH groups, whose contribution to the orientational polarization of cellulose is the largest, increases in the amorphous regions. An irradiation dose equal to about 3×10^5 Gy results in a decrease of the degree of crystallinity of the cellulose. The dielectric properties of cellulose reach their maximum value at a dose equal to about 10^6 Gy, which can be explained by an increase in the number of CH_2OH groups in both the crystalline and the amorphous regions. An irradiation dose above 10^6 Gy causes a rapid decrease in the dielectric properties of cellulose and of its degree of crystallinity. This result can be associated with a decrease in the number of CH_2OH groups due to their oxidation.

Analyses of Cole-Cole diagrams for oven-dry cellulose showed that an increase in the irradiation dose enlarges the distribution of the relaxation time. Figure 6.6 shows the influence of the irradiation dose on the dielectric loss factor of cellulose at various temperatures. It indicates clearly that unirradiated cellulose has its maximum value of dielectric loss factor at a frequency of 10^6 Hz, which at $t = 0\,°C$ is equal to about 0.22. At an irradiation dose of 97×10^4 Gy, the loss factor increases to 0.24 at $f = 10^6$ Hz. A further increase in the dose to 3.2×10^6 Gy diminishes the loss factor to 0.13 (at $t = 0\,°C$). Similar changes of the dielectric properties of wood occur upon irradiation with various doses of gamma-radiation as the dielectric properties of the wood are mainly determined

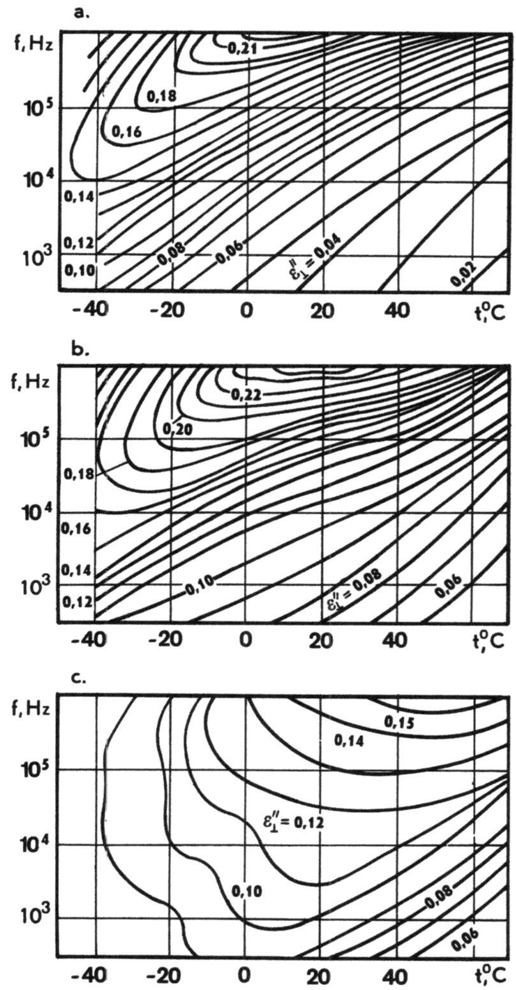

Fig. 6.6a–c. Dependence of the loss factor $\varepsilon''_{\perp}$ of cellulose irradiated with gamma radiation on frequency (f) and temperature (t), at density $\rho = 1.3$ g/cm^3 and moisture content $W = 0.3\%$ (Norimoto and Yamada 1975): **a** unirradiated cellulose; **b, c** irradiation dose 97×10^4 and 3.2×10^8 Gy

by the properties of cellulose. The dependence of the loss tangent value on the frequency for irradiated (with various doses), oven-dry birch wood measured with vector $\bar{E}$ oriented tangentially is indicated in Fig. 6.7 (Glukhov et al. 1973). At frequencies of up to 10^5 Hz, an irradiation dose below 5×10^5 Gy does not significantly change the value of $tg\delta$ of wood. At higher frequencies, with an increase in irradiation dose above 10^5 Gy, the maximum value of $tg\delta_{\perp}$ decreases, and its maximum shifts to the region of higher frequencies.

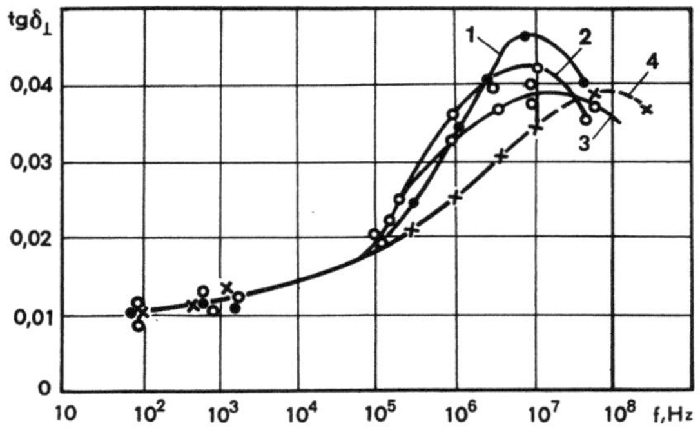

Fig. 6.7. Dependence of $tg\delta_\perp$ value of oven-dry birch wood irradiated with gamma radiation on frequency (Glukhov et al. 1973): *1* unirradiated wood; *2, 3, 4* wood irradiated with doses of 5×10^4, 10^5, and 5×10^5 Gy, respectively

From the above data, it can be concluded that at frequencies in the range of 40–10^6 Hz and at irradiation doses below 10^5 Gy, the dielectric properties of wood are virtually unchanged. With an increase of the dose up to 10^6 Gy, the DC increases 10%, and the value of the loss factor 20–30%. Any further increase of the dose upto 3.2×10^6 Gy diminishes the value of the DC 2–5% at frequencies up to 10^6 Hz, while the loss factor decreases 40% at frequencies of 10^4–10^6 Hz.

Chapter 7

Dielectric Properties of Bark

Bark is a more heterogeneous material than wood. Bark consists of two components, namely, inner bark (phloem) and outer bark, which differ from each other in layers, anatomical, chemical, and physical properties. The moisture content and density of these two components are quite different.

In elementary chemical composition, bark is similar to wood, but it possesses more mineral substances than wood, about 4%. For example, spruce bark contains: carbon 44%, oxygen 45.4%, hydrogen 6.4%, nitrogen 0.19%, potassium 0.37%, and phosphorus 0.76%. Organic components in bark are given in Table 7.1. Their relative amounts differ considerably from those in wood. Thus, bark contains much less cellulose. A substance such as suberin is present in bark and absent in wood. Therefore, the dielectric properties of oven-dry bark and wood should be very different.

Like wood, bark is a porous material. At a moisture content exceeding 30%, the dependence of the density of bark of some European tree species on the moisture content is similar to that of wood. For example, the ratio of the density of spruce bark to that of the wood at a moisture content of 55–120% is 0.93–1.14, for birch it is 0.89–1.11, and for pine it is 0.73–1.0. As the values of density and of moisture content are the principal factors determining the dielectric properties of wood, the dielectric properties of bark should be close to those of wood, provided their moisture content and density values are the same.

Table 7.1. Chemical composition of inner and outer bark in relation to the total mass (%) (Borovikov and Ugolev 1989)

Substances	Pine		Spruce		Birch	
	Inner	Outer	Inner	Outer	Inner	Outer
Cellulose	18.2	16.4	23.2	14.3	18.5[a]	3.4[a]
Lignin	17.1	43.6	15.6	27.4	20.3	1.3
Pentosans	12.1	6.8	9.7	7.1	20.2	1.1
Hexosans	16.3	6.0	9.3	7.7	–	–
Suberin	0	2.9	0	2.8	1.2	38.7
Extractives, soluble in:						
Water	20.8	14.2	33.1	27.9	–	–
Alcohol	3.9	3.5	1.7	2.6	13.7	5.6
Ether	–	–	–	–	1.7	38.1

[a] Including hexosans.

At low and high frequencies, the dielectric properties of bark have not been experimentally studied, and therefore the values of ε' and $tg\delta$ for bark (if needed for technical calculations) are usually assumed to be equal to those of the corresponding wood with an equal density value. With the development of microwave techniques and their application to moisture content measurements,

Table 7.2. Dielectric parameters of bark at microwave frequencies

Species	Moisture content (%)	Parameters	Temperature (°C) −30	−15	−5	0	20
At frequency 0.915 GHz							
Spruce	97	$\varepsilon'_{\perp}$	3.3	3.2	4.7	13	11.4
		$tg\delta_{\perp}$	0.17	0.21	0.22	0.20	0.24
		$\varepsilon'_{\parallel}$	4.9	4.8	7.1	23	20.5
		$tg\delta_{\parallel}$	0.21	0.25	0.26	0.24	0.28
Aspen	96	$\varepsilon'_{\perp}$	3.0	3.5	4.5	11.7	9.8
		$tg\delta_{\perp}$	0.15	0.21	0.22	0.22	0.23
		$\varepsilon'_{\parallel}$	4.4	5.3	6.7	21	17.6
		$tg\delta_{\parallel}$	0.17	0.25	0.26	0.26	0.27
Birch	69	$\varepsilon'_{\perp}$	3.2	3.4	4.4	8.3	7.3
		$tg\delta_{\perp}$	0.17	0.19	0.22	0.20	0.16
		$\varepsilon'_{\parallel}$	4.8	5.2	6.6	14.9	12.9
		$tg\delta_{\parallel}$	0.21	0.23	0.26	0.24	0.20
Pine	53	$\varepsilon'_{\perp}$	2.2	2.9	5.9	6.2	4.8
		$tg\delta_{\perp}$	0.14	0.19	0.23	0.24	0.18
		$\varepsilon'_{\parallel}$	3.4	4.3	8.9	11.2	8.6
		$tg\delta_{\parallel}$	0.17	0.23	0.27	0.28	0.22
At frequency 2.375 GHz							
Spruce	97	$\varepsilon'_{\perp}$	2.8	2.8	4.6	13.6	10.8
		$tg\delta_{\perp}$	0.16	0.19	0.24	0.18	0.23
		$\varepsilon'_{\parallel}$	4.2	4.2	7.0	24.5	19.4
		$tg\delta_{\parallel}$	0.19	0.23	0.28	0.22	0.27
Aspen	96	$\varepsilon'_{\perp}$	2.7	3.0	4.3	10.5	9.4
		$tg\delta_{\perp}$	0.15	0.17	0.25	0.25	0.22
		$\varepsilon'_{\parallel}$	4.1	4.4	6.5	18.9	16.8
		$tg\delta_{\parallel}$	0.19	0.21	0.29	0.29	0.26
Birch	69	$\varepsilon'_{\perp}$	2.9	3.2	3.7	6.7	6.4
		$tg\delta_{\perp}$	0.17	0.21	0.24	0.25	0.25
		$\varepsilon'_{\parallel}$	4.3	4.8	5.5	12.1	11.6
		$tg\delta_{\parallel}$	0.21	0.25	0.28	0.31	0.31
Pine	53	$\varepsilon'_{\perp}$	2.0	2.8	4.1	5.6	4.4
		$tg\delta_{\perp}$	0.15	0.22	0.25	0.23	0.18
		$\varepsilon'_{\parallel}$	3.0	4.2	6.1	10.0	8.0
		$tg\delta_{\parallel}$	0.19	0.26	0.31	0.27	0.22

to inspection of round timber to detect defects, to defrosting lumber, to preparation of lumber for debarking processes, and to application of debarking processes, the necessity of experimental determinations of the dielectric parameters of bark has arisen.

Measurements of the dielectric parameters of the bark of spruce, pine, birch, and aspen were carried out at frequencies of 0.915 and 2.375 GHz, which are allowed for industrial application. The measurements were performed in the cylindrical coaxial waveguide according to the procedure described in Sect. 3.3, and over a temperature range from − 30 to + 20 °C. The largest complication resides in the problem of preparation of the bark samples. The samples were cut out of the bark with a special appliance. They were made in the form of shavings of 5–8 mm thickness, which had to be tightly joined together to achieve a total thickness of 20 mm (for measuring at 2.375 GHz) and of 40 mm (for measuring at 0.915 GHz).

Measurements were carried out with bark samples with a moisture content of 53–97% prepared from the basal ends of fresh logs of spruce, pine, aspen, and birch trees. The dielectric constant (DC) and loss tangent were determined without separation of the outer and inner barks. The ratio of the phloem thickness to the total bark thickness was as follows: for pine 13–22%, for birch 85–86%, for aspen 74–97%, and for spruce 35–50%. The vector of electric field strength was oriented in the longitudinal-tangential plane, and the parameters ε' and $tg\delta$ in the longitudinal and the transverse directions were calculated with the use of relations determined for wood. The results are presented in Table 7.2. The values of ε' and $tg\delta$ of the bark depended on the temperature as in wood. With increasing temperature (starting from − 30 °C), the DC value gradually increased (Fig. 7.1), and at a temperature of about 0 °C it jumped sharply by 1.5–3 times depending on the moisture content. A rise of temperature above 0 °C caused a decrease of the DC value of the bark, which can be explained by a decrease of the free water DC with increasing temperature in the frequency range under consideration. The loss tangent value of bark has its peak value

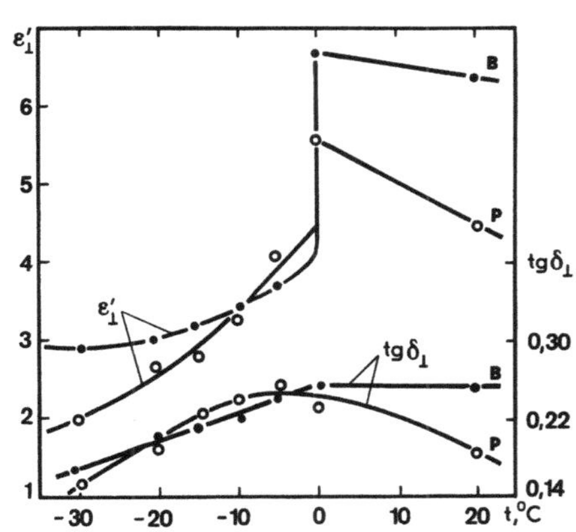

Fig. 7.1. Dependence of dielectric constant $\varepsilon'_\perp$ and loss tangent $tg\delta_\perp$ values of birch (B, $W = 69\%$) and pine bark (P, $W = 53\%$) on temperature (t), at frequency 2.375 GHz

about 0 °C. The average value of the variation coefficient of the bark ε' was equal to 19%, and that of $tg\delta$ 27%. The accuracy was 3.6% and 5.3%, respectively. The large variability range of the dielectric properties of bark is caused by the heterogeneity of its structure, the different ratios between inner and outer bark, and the nonuniformity in both moisture content and density.

The investigations showed that in the microwave range of frequencies at temperatures from 20 to 90 °C the values of the dielectric parameters of bark with a moisture content exceeding 30% do not differ from those of wood provided the moisture content and density values are equal and the conditions the same. Therefore, the values of ε' and $tg\delta$ determined for wood can be used in calculations of the influence of electromagnetic fields on bark.

Chapter 8

Dielectric Properties of Wood-Based Materials

8.1 Cellulose, Paper, and Board

Studies on the dielectric properties of cellulose intended for engineering purposes as well as of paper and board were primarily motivated by their use as electrically insulating materials in the cable industry and in the manufacture of capacitors. For the purpose of improving the insulating properties of cellulose and paper, they are impregnated with liquid insulating compositions that fill in the pores and form a dielectric material with the required dielectric properties. Introducing high frequency and microwave techniques into the manufacture of cellulose-based materials made it necessary to determine their dielectric properties within a wide range of frequencies. Studies of the dielectric properties of these materials in the high frequency and microwave ranges were also carried out as a response to the necessity of designing and developing instruments for measuring the moisture content of paper and board sheets, inspection of their density, mechanical and dielectric anisotropy, and airtightness, as well as for defect detection, and for nondamaging checking of strength.

The dielectric properties of paper and board are mainly determined by the properties of cellulose and its relative amount. The dielectric constant (DC) of cellulose with the vector $\bar{E}$ oriented perpendicular to the fiber direction in the frequency range from 10^2 to 10^{10} Hz changes within 3.6–7.1, and the loss tangent value accordingly changes within 0.018–0.080. The dielectric properties of cellulose, paper, board, and other cellulosics depend on their molecular constitution, physical structure, and chemical composition. Moreover, they are influenced by the following factors that characterize the properties of the material itself, namely, density, moisture content, tree species, anisotropy, quantity and composition of substances present, materials introduced and removed in the manufacture of cellulose and cellulose-based materials (glueing, filling, and coloring substances) as well as by external factors such as electromagnetic radiation frequency, the orientation of the electric field strength vector, and the temperature.

The variety of factors that act upon the dielectric properties of cellulose-based materials results in a wide variation in the dielectric parameters of these materials. It is difficult to estimate the influence of all these factors, and this is the reason why only principal factors, which are the ones most thoroughly studied will be described below. The dielectric properties of sheets of cellulose-based materials to a great extent depend on the orientation of the electric field strength vector $\bar{E}$. Because of this fact, our further examinations of dielectric properties of materials will be carried out relative to the orientation of the vector

$\bar{E}$ as follows:

Vector $\bar{E}$ is perpendicular to the surface of the sheet material; the designation is $\varepsilon'_{\perp}$ and $tg\delta_{\perp}$, respectively.
Vector $\bar{E}$ is parallel to the surface of the material and directed along the machine direction; the designation is ε'_{MD} and $tg\delta_{MD}$.
Vector $\bar{E}$ is parallel to the surface of the material and directed transverse to the machine direction (in the sheet plane); the designation is ε'_{CD} and $tg\delta_{CD}$.
Vector $\bar{E}$ is parallel to the sheet plane (without indication of either machine or cross direction); the designation is $\varepsilon'_{||}$ and $tg\delta_{||}$.

8.1.1 Influence of Density Value, Moisture Content, and Temperature on the Dielectric Properties of the Material

The methods of theoretical computation of the dielectric parameters of cellulose-based materials have been worked out only for particular cases and even in these as a rule empirical coefficients have been used. For practical purposes, experimental data referring to the dielectric characteristics of the materials are used.

Renne (1969), applying the theory of the series connection of pores and fibers in paper, suggested use of the following formulas for the calculation of ε' and $tg\delta$ of dry paper with a variable density. The DC value of paper can be found from

$$\varepsilon'_1 = \varepsilon'_c/[1 + g(\varepsilon'_c - 1)] \,, \tag{8.1}$$

$$g = 1 - \rho_1/\rho_c \,, \tag{8.2}$$

where

ε'_c = dielectric constant of the cellulose substance
g = relative volume of pores in the paper
ρ_1 and ρ_c = density of paper and of cellulose substance, respectively; the value of ρ_c is assumed to be equal to 1.55 g/cm^3.

The value of $tg\delta_1$ of the material is:

$$tg\delta_1 = tg\delta_c/[1 + g\varepsilon'_c/(1 - g)] \,, \tag{8.3}$$

where

$tg\delta_c$ = the loss tangent value of the cellulose.

Tentative estimations of the dielectric properties of paper and board in the oven-dry state may be performed with the use of formulas derived from Eq. (2.1) for ε' and Eq. (2.3) for $tg\delta$. Considering paper as a compound dielectric comprising two components, viz. cellulose and air, and assuming index "k" in Eq. (2.1) and (2.3) equal to 1/3 we find that

$$(\varepsilon')^{1/3} = V_1(\varepsilon'_c)^{1/3} + V_2(\varepsilon'_b)^{1/3} \,, \tag{8.4}$$

$$tg\delta = V_1(\varepsilon'_c)^{1/3} tg\delta_c/[V_1(\varepsilon'_c)^{1/3} + V_2(\varepsilon'_b)^{1/3}] \,, \tag{8.5}$$

where

V_1 and V_2 = cellulose and air content by volume
ε_c' and ε_b' = dielectric constants of cellulose and air
$tg\delta_c$ = loss tangent value of the cellulose substance.

The dielectric parameters of the wood cellulose (ε_c' and $tg\delta_c$) at various frequencies are given in Table 2.3 in Chap. 2.

A comparison of the calculated results for $\varepsilon_\perp'$ and $tg\delta_\perp$ values of paper and board using Eqs. (8.1) and (8.3) with dielectric parameter values of various kinds of paper and board obtained experimentally showed that the majority of experimental points plotted on a diagram turn out to be above the curves of $\varepsilon_\perp' = \varphi_1(f)$ and $tg\delta_\perp = \varphi_2(f)$. At the same time, these points lie below the curves $\varepsilon_\perp' = \varphi_3(f)$ and $tg\delta_\perp = \varphi_4(f)$ calculated with the aid of Eqs. (8.4) and (8.5). Thus, for example, according to Peskovets (1975) the DC value of dry board made from pine cellulose with an ash content equal to 0.28–0.43% and a density of 0.6 – 1.2 g/cm^3 in the frequency range from 10^6 to 4×10^7 Hz is 20–24% higher in comparison with the values computed using Eq. (8.1) and 14–26% lower than those determined with Eq. (8.4). At the same time, the values of $tg\delta_\perp$ are 1.4–3 times higher than the values found using Eq. (8.3) and are 1.4–1.8 times lower than the values calculated using Eq. (8.5).

Therefore, Eqs. (8.1), (8.3), (8.4), and (8.5) make it possible to determine zones on the diagrams $\varepsilon_\perp' = \varphi(f)$ and $tg\delta_\perp = \varphi(f)$ (Fig. 8.1) limiting the dielectric

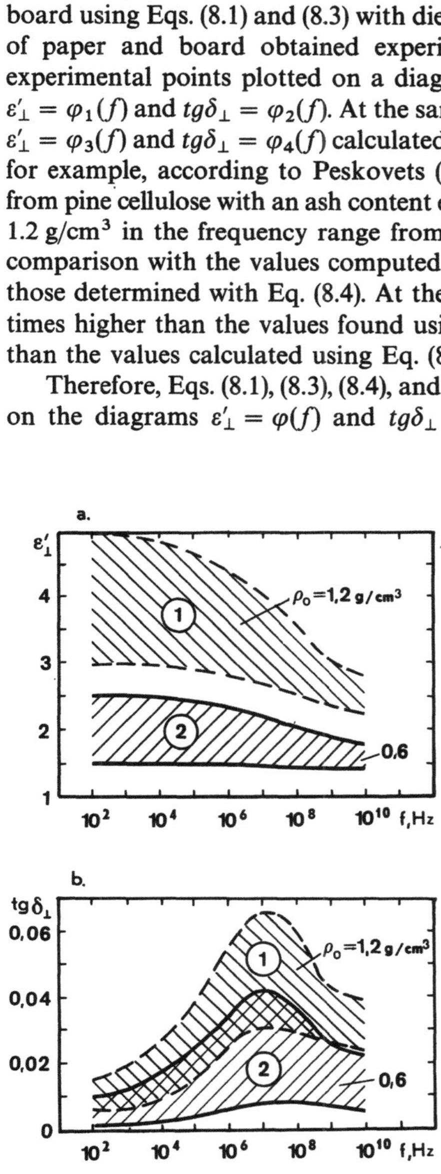

Fig. 8.1a, b. Frequency response of $\varepsilon_\perp'$ (**a**) and $tg\delta_\perp$ (**b**) of oven-dry paper and board of various brands, at 20–25 °C: zones 1 and 2 indicate limits of parameter values when the density $\rho_0 = 1.2$ g/cm^3 and 0.6 g/cm^3, respectively

parameter values of oven-dry paper (board) of variable density. Within zone 2 (Fig. 8.1) lie values of $\varepsilon'_\perp$ and $tg\delta_\perp$ with a density equal to 0.6 g/cm^3, and within zone 1 are the corresponding values when the density is equal to 1.2 g/cm^3. With increasing frequency, the dispersion of the DC values diminishes. The dispersion of $tg\delta_\perp$ increases with increasing frequency up to 10^7 Hz and then decreases. It should be noted that the zones of dielectric parameters have comparatively wide limits because of the wide variations in composition and structure of the materials studied. Table 8.1 shows the limiting values of the dielectric parameters of the most widely used kinds of paper and board. The data are related to oven-dry materials with a density from 0.6 to 1.2 g/cm^3 but do not take into account their composition and structure.

With an increase in the density of oven-dry paper (board), the values of $\varepsilon'_\perp$ and $tg\delta_\perp$ increase linearly (Fig. 8.2). As the board density increases from 0.6 to 1.2 g/cm^3 at a temperature of 25 °C and at a frequency of 10^6 Hz, the DC of the oven-dry board increases from 2 to 3.6; at a moisture content $W = 20\%$ from 2.5 to 4.6; at $W = 43\%$ from 4 to 8.2, i.e., by 1.8–2 times. At the same time, the value of $tg\delta_\perp$ increases as follows: at $W = 0\%$ from 0.022 to 0.031; at $W = 20\%$ from 0.26 to 0.36.

The influence of the moisture content on the dielectric properties of cellulose-based materials is the defining factor and the general nature of this influence is similar to that for wood. With an increase of moisture content in the region of monomolecular adsorption of water (up to 4.5–5.5%), the values of the DC and loss tangent of a board essentially do not change. With an increase in moisture content to the point at which polymolecular adsorption occurs [which according to Flate and Grunin (1973) in cellulose-based materials is completed when $W = 17$–19%], the value of the DC gradually increases. At a moisture content of 17–19%, the curve $\varepsilon' = \varphi(W)$ passes its inflection point, after which the slope of the curve greatly increases. This change can be explained by the appearance of capillary moisture in the material. On further increase of the

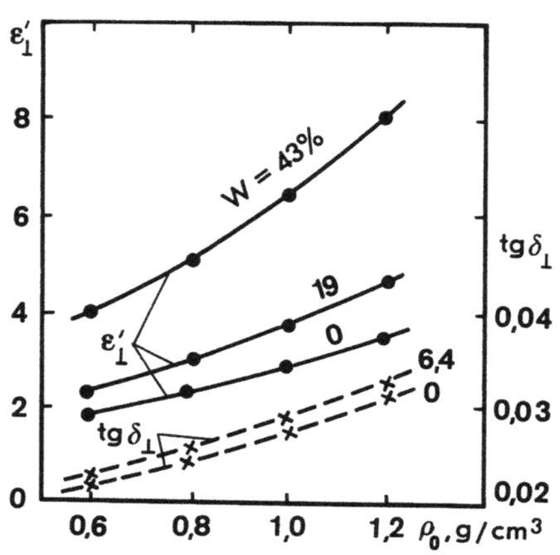

Fig. 8.2. Dependence of $\varepsilon'_\perp$ and $tg\delta_\perp$ values of board made from unbleached pulp upon density ρ_0 at a frequency of 10^6 Hz and temperature $t = 25$ °C: W moisture content (%) (Peskovets 1975)

Table 8.1. Limiting values of the dielectric properties of oven-dry paper and board at a density of $\rho_0 = 0.6\text{–}1.2\ \text{g/cm}^3$ and temperature $t = 20\text{–}25\,°\text{C}$, with vector $\bar{E}$ oriented perpendicular to the sheet surface

f (Hz)	Density (g/cm^3)							
	0.6		0.8		1.0		1.2	
	$\varepsilon'_\perp$	$tg\delta_\perp$	$\varepsilon'_\perp$	$tg\delta_\perp$	$\varepsilon'_\perp$	$tg\delta_\perp$	$\varepsilon'_\perp$	$tg\delta_\perp$
10^2	1.5–2.5	0.0012–0.010	1.8–3.3	0.0030–0.012	2.3–4.1	0.004–0.014	3.0–5.0	0.006–0.016
10^3	1.5–2.5	0.0018–0.012	1.8–3.3	0.0030–0.015	2.3–4.0	0.0046–0.017	2.9–4.9	0.007–0.019
10^4	1.5–2.5	0.0026–0.017	1.8–3.2	0.0044–0.021	2.3–4.0	0.0066–0.024	2.9–4.8	0.010–0.027
10^5	1.5–2.4	0.004–0.024	1.8–3.1	0.0067–0.031	2.2–3.8	0.010–0.035	2.9–4.6	0.016–0.039
10^6	1.5–2.3	0.006–0.034	1.8–3.0	0.010–0.043	2.2–3.6	0.015–0.049	2.8–4.3	0.023–0.055
10^7	1.5–2.2	0.008–0.042	1.8–2.8	0.014–0.053	2.1–3.4	0.020–0.061	2.7–4.0	0.030–0.068
10^8	1.4–2.1	0.008–0.036	1.7–2.6	0.013–0.046	2.1–3.1	0.019–0.053	2.6–3.7	0.028–0.060
10^9	1.4–1.8	0.007–0.025	1.6–2.2	0.011–0.032	1.9–2.5	0.016–0.037	2.3–2.9	0.023–0.042
10^{10}	1.4–1.8	0.007–0.023	1.6–2.1	0.011–0.030	1.9–2.4	0.016–0.035	2.3–2.8	0.022–0.039

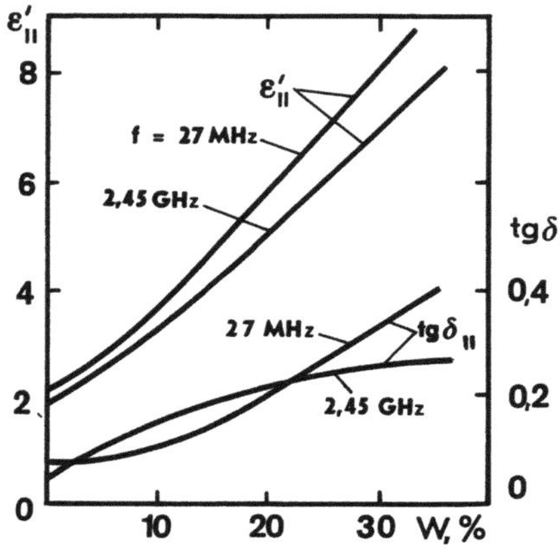

Fig. 8.3. Dependence of $\varepsilon'_{\|}$ and $tg\delta_{\|}$ of a board with density 0.54 g/cm^3 on moisture content (W, %) at $t = 19$–$23\,°C$ and frequencies 27 MHz and 2.45 GHz. Ash content is 3% (Metaxas and Driscoll 1974)

moisture content, the rise in the DC of the material caused by the water is different for each frequency range. With increasing frequency and at a constant moisture content, the value of the DC of the paper decreases. A sharp reduction in the DC value takes place at a frequency of about 10^{10} Hz, which is associated with a significant decrease in the DC value of water (Fig. 2.7).

The loss tangent value of paper and board at a high moisture content is also determined by the properties of water; the minimum value of $tg\delta$ is in the frequency range 10^7–10^8 Hz and maximum value, in the range 10^{10}–10^{11} Hz.

The nature of the changes in $\varepsilon'_{\|}$ and $tg\delta_{\|}$ of a board with a density of 0.54 g/cm^3 in relation to the moisture content at frequencies of 27 MHz and 2.45 GHz is shown in Fig. 8.3 (Metaxas and Driscoll 1974). The dielectric properties of the board in a frequency range from 10^6 to 4×10^7 is presented in Appendix 4.

Mikaelan (1978, 1982) studied the dielectric characteristics of cellulose-based materials in the microwave range. Figure 8.4 shows the dependence of $\varepsilon'_{\|}$ and $tg\delta_{\|}$ values of drawing paper (coniferous, sulfite-bleached, ash-free pulp) with a density of 0.8–0.9 g/cm^3, a mass of 160 g/m^2 at a wavelength in the range from 3 to 30 GHz. With increased moisture content from 2 to 60% and at a frequency of 30 GHz (wavelength equal to 1 cm), the value of $\varepsilon'_{\|}$ increases from 4.8 to 12, and at a frequency of 3 GHz from 3.3 to 17.9. The correlation between the DC and the wavelength is complicated. At 20 °C with increasing wavelength, starting from 1 cm, the value of $\varepsilon'_{\|}$ decreases, passes a distinct minimum value at $\lambda = 2$ cm (about 15 GHz) and then increases until an inflection point at $\lambda = 5$ cm (about 6 GHz). A further increase in the wavelength at moisture contents up to 50% causes a continuous diminishing of $\varepsilon'_{\|}$. At moisture contents above 50%, the value of $\varepsilon'_{\|}$ increases.

The loss tangent value of drawing paper (Fig. 8.4b) with a moisture content of 2% has a maximum of 0.41 at a wavelength of about 4 cm (7.5 GHz). With increasing moisture content, this maximum shifts to a higher frequency range, and its absolute value diminishes.

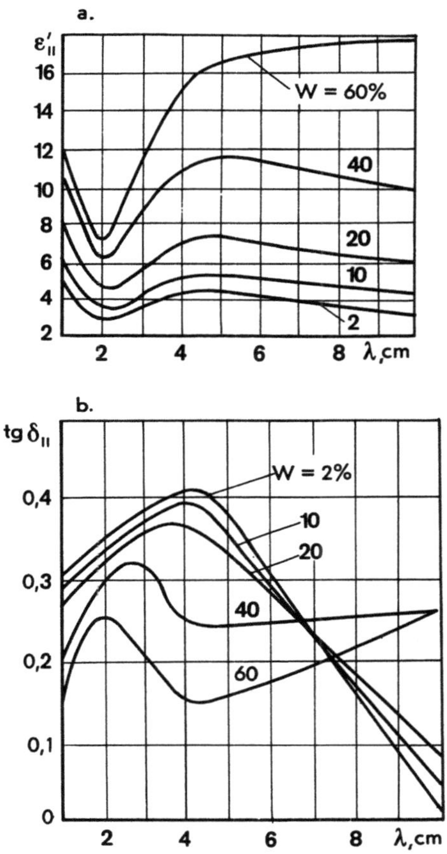

Fig. 8.4a, b. Dependence of $\varepsilon'_{\parallel}$ (**a**) and $tg\,\delta_{\parallel}$ (**b**) values of drawing paper in the microwave frequency range on wavelength, at the moisture content $W = 2$–60%, and $t = 20\,°C$ (sulfite softwood pulp, bleached, ash-free, with a density of 0.8–0.9 g/cm^3, mass 160 g/m^2) (Mikaelan 1978)

Figure 8.5 shows the dependence of the dielectric characteristics of a special brand of paper [sulfate (kraft) prehydrolysis pulp] with a density of 0.8 g/cm^3 upon the moisture content and temperature at a frequency of 10 GHz. With increasing moisture content, the DC increases parabolically. An increase of moisture content from 0 to 60% causes the DC to increase 3.5–7.5 times, depending on the temperature. The change in temperature from 30 to 90 °C increases the DC of oven-dry paper from 2.4 to 6, and at a moisture content of 60%, this increase is from 18 to 21. The loss tangent value of this paper at a frequency of 10 GHz increases with increasing moisture content, starting from 0%, and reaches its maximum value at $t = 20\,°C$ and a moisture content of about 20%. With increasing temperature, the maximum value of the loss tangent sinks and its maximum shifts toward lower moisture contents (Fig. 8.5b). An increase of temperature from 20 to 90 °C at a moisture content equal to 20% causes a decrease of $tg\delta$ from 0.56 to 0.32, whereas at $W = 60\%$, this reduction of $tg\delta$ is from 0.41 to 0.21.

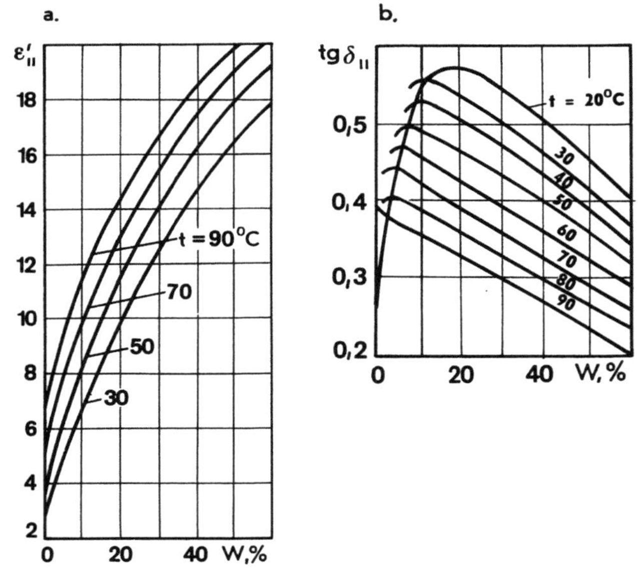

Fig. 8.5a, b. Dependence of $\varepsilon'_{\parallel}$ (**a**) and $tg\delta_{\parallel}$ (**b**) values of a special brand of paper on the moisture content at $t = 20$–$90\,°C$ and at a frequency of 10 GHz (sulfate prehydrolysis pulp cellulose, with iron content up to 40 mg/kg, calcium up to 60 mg/kg, ash-free), density 0.8 g/cm³, mass 30 g/m², thickness 0.040–0.042 mm (Mikaelan 1982)

The dielectric properties of moist drawing paper as measured in the microwave frequency range are given in Table 8.2. Appendix 5 shows the dielectric parameters of various kinds of cellulose and paper at $t = 30$–$90\,°C$ and a moisture content of $W = 0$–80%, as measured at 10 GHz.

The characteristics of the changes in dielectric properties of oven-dry paper (board) are similar to those of oven-dry wood. With increasing temperature, the DC increases continuously at any frequency. Depending on the frequency, the maximum values of the loss tangent are at the following temperatures: at a frequency of 10^3 Hz, the maximum value of $tg\delta$ is at $-80\,°C$; at 10 Hz, at $-60\,°C$, and at 10^5 Hz, at about $-35\,°C$ (Breitweit et al. 1970). At higher frequencies, the maximum of $tg\delta$ shifts in the direction of positive temperatures. The average temperature coefficient of the DC of dry writing paper, $k^t_{\varepsilon'}$, in the temperature range from 20 to 80 °C is correlated with the frequency as follows:

$$f = 10^2\text{–}10^4\ \text{Hz},\quad k^t_{\varepsilon'} = 12\times10^{-4}\ \text{to}\ 15\times10^{-4}\,\text{K}^{-1}$$

$$f = 10^5\text{–}10^8\ \text{Hz},\quad k^t_{\varepsilon'} = 17\times10^{-4}\ \text{to}\ 20\times10^{-4}\,\text{K}^{-1}$$

$$f = 3\times10^8 - 10^{10}\ \text{Hz},\quad k^t_{\varepsilon'} = 15\times10^{-4}\ \text{to}\ 16\times10^{-4}\,\text{K}^{-1}$$

The average temperature coefficient of the loss tangent of dry paper $k^t_{tg\delta}$ in the temperature range from 20 to 80 °C in a frequency range of 10^3–10^7 Hz is

Table 8.2. Dielectric properties ($\varepsilon'_{\|}$ and $tg\delta_{\|}$) of moist paper with a density $\rho_0 = 0.8$–$0.9\ g/cm^3$ made from coniferous, ash-free, sulfite pulp, in a microwave frequency range, at $t = 20\,°C$ (Mikaelan 1982)

f (GHz)	Moisture content (%)									
	2		10		20		40		60	
	$\varepsilon'_{\|}$	$tg\delta_{\|}$	$\varepsilon'_{\|}$	$tg\delta_{\|}$	$\varepsilon'_{\|}$	$tg\delta_{\|}$	$\varepsilon'_{\|}$	$tg\delta_{\|}$	$\varepsilon'_{\|}$	$tg\delta_{\|}$
3	3.3	0.03	4.2	0.056	6.0	0.093	9.8	0.26	17.9	0.25
5	4.1	0.31	5.2	0.29	7.1	0.28	11.4	0.25	17.2	0.17
7.5	4.4	0.42	5.3	0.40	7.1	0.36	11.0	0.25	15.3	0.15
10	3.7	0.39	4.4	0.38	5.8	0.36	8.6	0.32	11.3	0.21
15	2.9	0.35	3.8	0.33	4.5	0.32	6.2	0.30	7.1	0.25
30	4.8	0.31	6.0	0.29	8.0	0.27	10.6	0.20	12.0	0.15

Table 8.3. Values of temperature coefficient of loss tangent of dry paper $k^t_{tg\delta}$

Frequency (Hz)	10^3	10^4	10^5	10^6	10^7	3×10^8	10^9	10^{10}
$k^t_{tg\delta} \times 10^4\ (K^{-1})$	− 7	− 80	− 100	− 70	− 40	+ 16	+ 75	+ 185

negative, but in the range from 3×10^8 to 10^{10} Hz it is positive; its values are given in Table 8.3.

The dielectric properties in the temperature range from 20 to 80 °C can be calculated if the values of DC and $tg\delta$ for paper at 20 °C are known, by using Eqs. (1.40) and (1.41) and the above coefficients of $k^t_{\varepsilon'}$ and $k^t_{tg\delta}$.

An increase of temperature of moist cellulose-based materials in the range from 0 to 95 °C causes an increase of the DC in the entire frequency range from 0 to 10^{11} Hz. However, the extent of the influence of temperature at any of the specified frequencies is different, as is also characteristic for wood. The loss tangent value of moist materials increases with increasing temperature from 0 to 95 °C in the frequency range from 20 to 10^8 Hz, which is caused by the positive temperature coefficient of water in this range (Fig. 2.7). At frequencies exceeding 10^8 Hz, the value of $tg\delta$ for water decreases with increasing temperature. This is the reason why the loss tangent value of moist cellulose-based materials is reduced at frequencies above 10^8 Hz. The literature on the subject does not give any data on the temperature coefficients of the DC and $tg\delta$ values of moist paper and board, and therefore as an approximate reference point, the values of $k^t_{\varepsilon'}$ and $k^t_{tg\delta}$ for moist wood (Chap. 5) can be used.

The dielectric properties of thoroughly dried cellulose-based materials do not depend on the electric field strength value provided that this value is kept within limits that exclude any ionization processes in the pores of the materials.

8.1.2 Influence of the Structure of Paper and Board on Their Dielectric Properties

In the three-dimensional structure of paper and board, most fibers are oriented parallel to the surface. However, some of the fibers are arranged at some angle to the surface. It is impossible to give a quantitative or even qualitative estimation of the influence of these fibers on the dielectric properties of the material. Therefore, as a rule this heterogeneity of the material is neglected. The nonuniformity of the density distribution along the sheet is also neglected.

The orientation of fibers in the horizontal plane of the sheet is a result of processing technology and of the thickness of the material. This orientation can vary from mainly the machine direction to an orientation in any direction.

According to the *Papermaker's Guide* (Alekseev et al. 1965) the most widely used paper production processes result in the following values for the quantity of fibers oriented mainly in the machine direction and of those mainly in the transverse direction: On the wire side of paper, this value is 3.2 to 1, and on the opposite side, it is equal to 1.5 to 1. Considering the average value of this relation for the whole thickness of the paper sheet, we can assume that paper consists of 70% of fibers oriented along the sheet and 30% in the transverse orientation. These data, however, do not allow any conclusions to be drawn regarding the relation between the dielectric properties of paper in the machine and the transverse directions.

Indirectly, the fiber orientation is characterized by the difference in mechanical strength in the longitudinal (machine) and the transverse directions. According to Flate (1986), the ratio of the breaking strength of paper in the longitudinal and the transverse directions varies from 0.8 to 1 to 7 to 1.

The strength anisotropy of a material indirectly characterizes the anisotropy of its dielectric properties. Assuming that at an equal mechanical strength in the longitudinal and the transverse directions the fibers are oriented uniformly in all directions, the dielectric parameters should be equal to the $\bar{E}$ orientation in the machine and in the transverse directions. In all other cases, these values differ considerably from one another. Thus, according to Osaki (1989) at a ratio of the material strength in the machine and the transverse directions of 4.9, the relation $\varepsilon'_{MD}/\varepsilon'_{CD}$ at a frequency of $f = 4$ GHz, $W = 9\%$, and $t = 20\,°\mathrm{C}$ was 1.09, and the relation $tg\delta_{MD}/tg\delta_{CD}$ was 1.3.

Analysis of experimental data obtained by Metaxas and Driscoll (1974) on the dielectric parameters of paper and board at $f = 27$ MHz, $t = 19$–$23\,°\mathrm{C}$, and with the vector $\bar{E}$ oriented perpendicular to the sheet surface and in the machine direction showed that the relation $\varepsilon'_{MD}/\varepsilon'_{\perp}$ is equal to 1.5 at $W = 0$–5%, 1.8–2.2 at $W = 10$–20%, and 2.5–2.6 at $W = 25\%$–30%. In comparison with the relation $k_1 = \varepsilon'_{\parallel}/\varepsilon'_{\perp}$ for coniferous wood, the relation $\varepsilon'_{MD}/\varepsilon'_{\perp}$ for paper and board with a moisture content up to 30% is only 5–20% higher than that of wood.

The relation $tg\delta_{MD}/tg\delta_{\perp}$ under the same conditions for paper with a density of 0.61 g/cm^3 and at $W = 5$–30% was 1.4–1.6, while for a board with a density of 0.54 g/cm^3, it was 2.0–2.5. The conclusion may therefore be drawn that in the range of radio frequencies the dielectric properties of paper and board with a moisture content up to 30% in the machine direction are 1.4–2.5 times higher than those in the direction perpendicular to the sheet surface. With increasing frequency, the difference between the values for the dielectric properties of wood with the vector $\bar{E}$ orientation along and across the fiber direction diminishes. Proceeding from this, it can be concluded that at frequencies above 10^8 Hz, the difference between the dielectric properties of cellulose-based materials, which is expressed by the relations $\varepsilon'_{||}/\varepsilon'_{\perp}$ and $tg\delta_{||}/tg\delta_{\perp}$, should diminish as well.

The literature on the subject gives values for the dielectric parameters of cellulose-based materials without indicating the orientation of the electric field strength vector in the sheet surface plane, which introduces serious errors into the calculations relating to the interactions between the electromagnetic fields and the materials under consideration.

During the process of pulp beating, its fibers shorten, and their external surface enlarges. Experiments carried out at frequencies from 5×10^4 to 1.5×10^7 Hz showed that at a moisture content below 12% the DC of the final board product is virtually independent of the extent of grinding. This can be explained by the low mobility of the molecules of water adsorbed on the surface of the fibers. The mobility of the molecules of adsorbed water is nearly independent of the extent of grinding (Flate and Grunin 1973). According to data by Peskovets (1975) at moisture contents from 16 to 70% and an increase in the extent of grinding by 4.4 times the pulp DC value diminishes by 13–15%.

8.1.3 Influence of Organic and Mineral Inclusions on the Dielectric Properties of Cellulose-Based Materials

There are differences of opinion concerning the influence of hemicelluloses and lignin on the dielectric properties of cellulose-based materials. Some investigators believe that the elimination of lignin results in a comparatively small decrease in $tg\delta$, namely, only by 9–10%. On total elimination of the hemicelluloses, for example, in the production of pure cellulose, the value of $tg\delta$ increases to 42%. According to data obtained by other investigators, a decrease of the pentosan content to 5–6% causes a notable decrease in the $tg\delta$ at temperatures from 20 to 120 °C. Lowering the content of lignin to within 2–6% in the samples did not result in any changes in $tg\delta$. According to Renne (1969), an increase of the α-cellulose content in paper results in an increase in the DC. The $tg\delta$ also varies with temperature. At temperatures below 45 °C, when the relative content of α-cellulose is increased from 89 to 99%, the value of $tg\delta$ decreases, while at temperatures above 45 °C, it increases.

A comparison between the dielectric properties of cellulose and lignin (Tables 2.3, 2.4) shows that a decrease in lignin content causes an increase of the DC of the material, as ε'_l of lignin at $t = 20$ °C and at frequencies from 50 to 10^7 Hz is 1.5–1.7 times lower than the ε'_c of cellulose. The loss tangent value of cellulose under the same conditions is 4–5 times higher than that of lignin. This is the reason why the influence of the $tg\delta_l$ value of lignin on the loss tangent of cellulose-based materials is so insignificant that it is difficult to assess it by measurements. However, this reasoning does not take into account the fact that changes in the structure of the material can significantly influence its dielectric properties.

A study of the correlation between the degree of delignification of pine wood in the process of cellulose production and its loss tangent value were carried out by Milov et al. (1974). They investigated samples of cellulose intended for industrial purposes with a lignin content from 3.2 to 16.1% and with a polymerization degree of 2200–2600. The pentosan contents of the samples were 8.5–9.7%. The value of $tg\delta$ obtained showed that within t = 20–90 °C a decrease in the lignin content was associated with a decrease of $tg\delta$. However, at increasing temperatures within the range 90–120 °C, the lowest values of the dielectric loss tangent were exhibited by samples with a lesser degree of delignification. The increase in $tg\delta$ in the most delignified samples at t = 100–120 °C can be explained by active oxidation processes which take place in the samples. In samples with a lignin content of 3.2%, the polymerization degree of cellulose diminished from 1380 to 890, while the carboxyl group content increased from 0.11 to 0.17%.

The influence of the pentosan (xylan) content on $tg\delta$ at high temperatures differs from that at low temperatures. At low temperatures, its influence diminishes with a decrease in the hemicellulose content. In contrast, samples with a higher content of hemicelluloses at temperatures of 100–120 °C have lower values of $tg\delta$ (Milov et al. 1974).

The influence of the pentosan content on $tg\delta$ of capacitor papers of different compositions was studied by Ivanov (1984). The data obtained are given in Table 8.4. On the basis of these data, Ivanov came to the conclusion that at low frequencies in the temperature range from 20 to 120 °C, oven-dry paper has the

Table 8.4. The loss tangent value $tg\delta_\perp$ of oven-dry capacitor paper of varying composition (density $\rho_0 = 1\ g/cm^3$, ash content 0.1–0.23%, high degree of grinding, frequency 50 Hz) (Ivanov 1984)

Percentage by mass			$tg\,\delta_\perp \times 10^4$		
Pentosans	α-Cellulose	Lignin	40 °C	90 °C	120 °C
2.5	95.2	2.3	20	30	46
4.0	93.3	2.7	12	25	35
4.4	93.0	2.6	14	18	31
4.8	92.7	2.5	15	18	26
6.2	90.8	3.0	16	18	20
7.0	90.0	3.2	18	15	22
9.4	87.3	3.3	18.	20	36

lowest $tg\delta$ at the following composition (percentage by mass): α-cellulose 90–93%, pentosans 5–7%, lignin 2.5–3%.

Paper manufactured from refined cellulose with an increased content of α-cellulose has a small absolute value of $tg\delta$; its changes with increasing temperature are comparatively small, provided the paper does not contain added minerals.

The influence of the tree species from which the cellulose is manufactured on the $tg\delta$ of the paper can be gleened from the data reported by Breitweit et al. (1970). At $t = 100\,°C$, the $tg\delta$ of dry cable-insulating paper made from cellulose from the following tree species was equal: pine 0.0017, spruce 0.0020, aspen 0.0018, and birch 0.0024.

The ash content is the most important characteristic of cellulose intended for the manufacture of electrically insulating materials. The presence of ash increases the dielectric losses of a material. The nature of the mineral components and their composition are also important. The content and composition of the ash ingredients are influenced by various factors. One of them is the kind of wood and the place of its growth. The general content of alkali metals, calcium, and magnesium in the heartwood and in the sapwood is similar, but their ratio is different (sapwood contains proportionally more alkali metals).

The quantity of ash in a kraft pulp is higher than that in wood. This is explained by contamination of the pulp in the pulping process with agents and salts that depend on the hardness of the water used. A considerable proportion of the mineral substances with a low solubility in water is present in the form of inclusions and is weakly bonded to the fibers. Another part of the ash is more firmly bound to cellulose, and it is very difficult to produce cellulose with a low (0.07–0.1%) ash content.

Ash ingredients in cellulose intended for industrial purposes is present in the following forms:

Mechanically bound ash containing mineral substances with a low solubility, which are detained on the surface of or inside the fibers

Exchangeable ash, i.e., metal cations bound with acidic groups in substances consisting of cellulose-based materials that are capable of cationic exchange reactions

Nonreactive ash made up of mineral substances that cannot be removed after treatment with weak acid but that do not take part in the exchange reactions.

Inorganic admixtures of variable proportions and composition are the cause of the increase in the dielectric characteristics of cellulose-based materials. A similar influence on the dielectric properties is exerted by chlorides, sulfates, and sulfides.

For electrical insulating papers and boards, the principal factor of the two dielectric parameters ε' and $tg\delta$ is the loss tangent value. This is the reason why the loss tangent has been investigated more than the DC.

Cations of monovalent metals such as potassium, sodium, and lithium affect $tg\delta$ more than the cations of the trivalent aluminum. The least dielectric losses are suffered by paper samples saturated with hydrogen, as well as with cations of divalent metals such as magnesium, calcium, and barium (Milov and Kitajeva 1959, Milov et al. 1974). Ions of sodium and mercury at temperatures below 80 °C increase the dielectric parameters. Ions of silver and iron at a temperature of 80 °C and above sharply increase them. At a temperature of 100 °C, ions of copper, zinc, and aluminum increase the dielectric properties. Ions of calcium and tin increase them at temperatures above 120 °C.

The conductivity of cellulose intended for industrial purposes is determined by the motion of ions that are included in the polymer but particularly by the movements of loosely bound ions of additives. The loss of energy caused by the conductivity phenomenon becomes more intense in cellulose upon a rise of temperature, which enhances the mobility of the ions. The energy losses created by the dipole polarization exceed the losses caused by conductivity, a fact that is extremely notable at high frequencies.

The quality of industrial water is of particular importance in the manufacture of electrically insulating cellulose and paper as it profoundly affects the content and composition of the ash ingredients. The increase of ε' and $tg\delta$ of paper is caused by a high content of the hygroscopic salts of calcium and magnesium in the paper as well as of sulfate anions which are present in the industrial water.

Dolgaleva (1958) and Kitajeva (1960) arranged cations in the following order according to their ability to increase the DC and loss tangent values of paper: $Na^+ > Li^+ > K^+ > Al^{3+} > NH_4^+ > Ca^{2+}$.

De-ashing of cellulose by electrodialysis (Kitajeva and Korolev 1957) to 0.12–0.04% causes a decrease $tg\delta$ and changes its temperature curve (Fig. 8.6). Thorough de-ashing of cellulose makes it possible to extend the admissible temperature limits in the usage of paper-based dielectric materials.

The purity of the water used for the manufacture of electrically insulating cellulose profoundly influences its properties. The use of water with an electric conductivity of 2×10^{-6} ohm^{-1} cm^{-1} for washing the cellulose decreases the

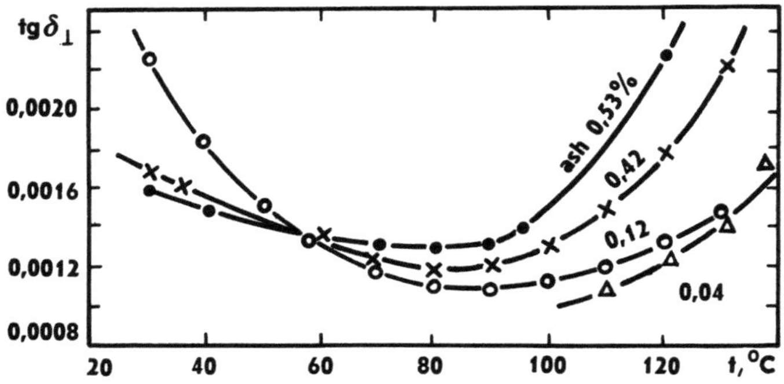

Fig. 8.6. Influence of the cellulose de-ashing degree on the $tg\delta_{\perp}$ value (Milov et al. 1974)

$tg\delta$ value of the insulating paper at 80–100 °C by 15%–35% in comparison with paper washed with water with an electric conductivity equal to 3×10^{-4} ohm^{-1} cm^{-1}. Cellulose washed with distilled water has the lowest loss tangent value.

8.2 Fiberboards

The dielectric parameters of fiberboards have been studied for two main reasons. First, fiberboards can be used as insulators and protective guards in some unimportant structures. The other reason is that electromagnetic fields are being used in the production process of fiberboards. The knowledge of their dielectric properties is also used for quality inspection purposes (checking thickness, moisture content, density values, and roughness of the surface).

Three methods are currently used in the manufacture of fiberboards, namely, the wet process, the semidry process, and the dry process. In the production of fiberboards, the wood is subjected to a defiberization process and to thermal and moisture treatment. Following this, binding and other substances are added to achieve the required properties of the fiberboards, such as water resistance, temperature resistance, and improved life span. All these processes affect the dielectric properties of the fiberboards. Noncompressed fiberboards have a density of 0.1–0.4 g/cm^3. Compressed fiberboards are manufactured in three grades according to their density: fiberboards of medium density of 0.4–0.8 g/cm^3, hardboards with a density of 0.8–1.2 g/cm^3, and compressed hardboards with a density of 1.25–1.45 g/cm^3. The dielectric properties of the three grades of fiberboards vary in proportion to their densities.

The relationship between the dielectric properties of fiberboards on their density, moisture content, tree species, temperature, frequency range, and vector $\bar{E}$ orientation is similar to those of wood. Particular features of fiberboards have their source in the extent of wood grinding, in the structure of the fiberboards as determined by the production process, and in the character of the substances introduced in the fiberboards in their manufacture.

Relations between the dielectric properties of oven-dry fiberboards and wood can be estimated by their comparison at various frequencies and at a temperature of 25 °C. It can be assumed that the fibers in the fiberboards are arranged parallel to the sheet plane. When vector $\bar{E}$ is positioned perpendicular to the sheet plane, it is also oriented perpendicular to the fiber, corresponding to the same direction in wood (perpendicular to the fibers). This makes it possible to compare the dielectric properties of wood and fiberboards under approximately equal conditions.

Comparison of the $\varepsilon'_{\perp f}$ values of oven-dry fiberboards made by the wet, semi-dry, and dry production processes at $t = 20–25$ °C with those of $\varepsilon'_{\perp}$ of oven-dry wood showed that in the frequency range from 10^2 to 10^8 Hz, the difference between these values lies between -10 and $+7$%, and that with increasing frequency, this difference diminishes. Hence, the conclusion can be drawn that the DC of oven-dry fiberboards with a vector $\bar{E}$ orientation perpendicular to the sheet plane for practical calculations can be assumed to be equal to $\varepsilon'_\perp$ of wood of similar density. The values of $\varepsilon'_\perp$ for wood are given in Appendix 1.

The loss tangent value of oven-dry fiberboards $tg\delta_{\perp f}$ with vector $\bar{E}$ oriented perpendicular to the sheet plane is always higher than the $tg\delta_\perp$ value of oven-dry wood of the same density. The empirical values of the relation $tg\delta_{\perp f}/tg\delta_\perp$ at $t = 25\,°C$ relative to the frequency are equal:

$f = 10^2$ Hz, 1.1–2.4

$f = 10^3$ Hz, 1.1–1.4

$f = 10^4$–10^7 Hz, 1.2–1.4

$f = 5 \times 10^7$ Hz, 1.7–1.8

The maximum value of $tg\delta_{\perp f}$ is shifted to a higher frequency range in comparison with the maximum value of $tg\delta_\perp$ of wood. The values of the loss tangent of oven-dry fiberboards are higher than those of wood. This fact can be explained by the different structure of fiberboards and also by the presence of various substances introduced into the fiberboards. The presence of polar groups such as OH and CH_2OH in the oven-dry fiberboards causes a dipole-radical polarization, and the presence of free ions and ions of various additives adsorbed by the molecules of the substances forming the wood governs their conductivity.

The dielectric properties of hardboards in the frequency range from 20 to 5×10^7 Hz have been thoroughly explored by James (1975). In the oven-dry state, hardboards with a density of 0.9–1.0 g/cm^3, at $t = 20\,°C$, and with vector $\bar{E}$ oriented perpendicular to the plane of the boards showed the following frequency response of $\varepsilon'_{\perp f}$:

$f = 20$ Hz, the DC was within 3.3–4.1

$f = 10^6$ Hz, 3.1 to 3.5

$f = 5 \times 10^7$ Hz, 2.9–3.2

With increasing frequency, the DC of oven-dry hardboards decreases continuously. A rise in temperature from -20 C to $+90\,°C$ increases the DC value at a frequency equal to 20 Hz from 3.3–4.0 to 4.4–4.5, and at a frequency of 5×10^7, from 2.6–2.9 to 3.2–3.6, i.e., by 13–33%.

The loss tangent value of these oven-dry boards increases at $t = 25\,°C$ from 0.017–0.021 to 0.071–0.079 with an increase in frequency from 20 to 5×10^7 Hz. The correlation of the loss tangent value with the temperature at various frequencies is complicated in character. At frequencies of 20–10^3 Hz, the value of $tg\delta_{\perp f}$ has its minimum at a temperature of about 20 °C. At higher frequencies (10^4–10^6 Hz), the value of $tg\delta_{\perp f}$ decreases with increasing temperature from $-20\,°C$ to 90 °C, and at frequencies of 10^7–10^8 Hz, it increases from 0.05 to 0.08–0.09.

Oven-dry fiberboards manufactured by wet or dry processes of compression do not reveal any notable difference in the loss tangent values.

Fiberboards are regarded as a hygroscopic material, and they may have a moisture content up to 30–35% depending on the humidity of the environment. The dielectric properties of the boards can vary accordingly. At low frequencies, an increase of the moisture content from 0 to 30–35% (at $t = 25\,°C$) causes an increase of the DC at $f = 20$ Hz from 3.5–4 to 3500–33 000, at frequencies of 10^2 Hz up to 800–5700, and at frequencies of 10^3 Hz upto 170–690 (James 1975). With increasing moisture content, the DC at frequencies of 10^4–10^5 Hz rises sharply, following a parabolic curve, but at frequencies above 10^6 Hz, this dependence on the moisture content is nearly linear. At a frequency of 10^7 Hz, an increase in the moisture content up to 30% results in a linear increase in the DC from 3.5 to 9.5. With increasing frequency, (Fig. 8.7a),

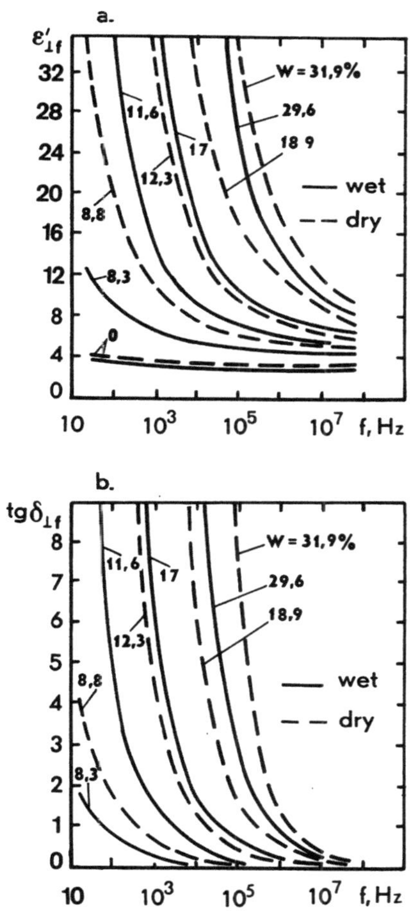

Fig. 8.7a, b. Frequency response of $\varepsilon_{\perp f}$ (**a**) and $tg\delta_{\perp f}$ (**b**) values of hardboards of 0.9–1.0 g/cm³ density, made by wet and dry processes: *W* moisture content (%) (James 1975)

the DC of moist hardboards decreases, provided the temperature does not change.

An increase in temperature from 25 to 90 °C (at $W = 12\%$) results in the following increase in the DC: at frequencies of 10^2–10^3 Hz from 15–70 to 30–530, and at frequencies of 10^6–10^7 Hz from 5.5–8.0 to 6.3–14. The DC of hardboards manufactured by the dry process is higher than that of boards made by the wet process, assuming that all other conditions are the same.

With increasing moisture content of the boards, the loss tangent value increases (at low frequencies equal to 20–10^3 Hz and $t = 25$ °C) from 0.02–0.04 (at $W = 0\%$) to 70–140 (at $W = 30$–35%). At frequencies of 10^6–10^7 Hz, the value of $tg\delta_{\perp f}$ under the same conditions increases from 0.05–0.06 to 0.3–2.4.

The frequency response of $tg\delta_{\perp f}$ of moist hardboards at $t = 25$ °C in the frequency range from 20 to 5×10^7 Hz is given in Fig. 8.7b. With increasing temperatures from $-$ 20 to 90 °C, at a frequency of 10^6 Hz, and with a moisture content $W = 0\%$, the value of $tg\delta_{\perp f}$ decreases from 0.056 to 0.028. At a moisture content $W = 8.3\%$, it decreases with an increase in temperature from 0.085 to its minimum value of 0.043 at 45 °C and then increases to 0.068 at 90 °C. At a moisture content above 10%, it increases with increasing temperature, following a parabolic curve. At a moisture content $W = 8$–12%, $t = 25$ °C, and under otherwise equal conditions, the loss tangent of hardboards made by the dry process is higher than that of hardboards manufactured by the wet process. At a moisture content of 8–12% and at a temperature of 25 °C, the $tg\delta_{\perp f}$ is 2–4 times higher at a frequency of 10^6 Hz, and at frequencies of 10^7–5×10^7 Hz, it is 1.1–1.6 times higher.

Hardboards made from oak wood by the dry process have a DC $\varepsilon'_{\perp f}$ (at a moisture content of 9–12%) that is 1.3–3.0 times lower than that of boards made from softwood at frequencies of 10^2–10^4 Hz; at frequencies of 10^5–10^7 Hz, it is 1.05–1.3 times lower. The loss tangent value of hardboards made from oak wood is lower than that of those made from softwood: at frequencies of 10^2–10^5 Hz, it is 2–3.5 times lower, and at frequencies of 10^6–10^7 Hz, it is 1.1–1.5 times lower.

The high values of the dielectric properties of moist fiberboards can be explained by a considerable increase of their conductivity at low frequencies and by an increase of dipole and interfacial polarization at high frequencies.

Experimental data relating to the dielectric properties of fiberboards in cases when the vector $\bar{E}$ of the electric field is oriented parallel to the sheet surface cannot be found in the literature. Assuming that the wood fibers in the fiberboards are equally oriented in the sheet plane in all directions, it becomes possible to calculate the dielectric parameters of the material with the orientation of vector $\bar{E}$ parallel to the sheet surface by using the empirical relations $k_1 = \varepsilon'_{||}/\varepsilon'_{\perp}$ and $k_2 = tg\delta_{||}/tg\delta_{\perp}$ valid for wood. These dielectric parameters should then be:

$$\varepsilon'_{||f} = (1 + k_1)\varepsilon'_{\perp f}/2 \,, \tag{8.6}$$

$$tg\delta_{||f} = (1 + k_2)tg\delta_{\perp f}/2 \,. \tag{8.7}$$

The values k_1 and k_2 for oven-dry wood are to be taken from Table 4.3 and for moist wood, from Tables 5.5 and 5.17.

The dielectric parameters of hardboards in the frequency range from 20 to 5×10^7 Hz at various temperatures are given in Appendix 6.

8.3 Particleboards

The great variety of properties that are characteristic of particleboards are caused by various factors acting upon this material during the production process. First of all, the properties depend on the tree species from which the boards are manufactured, and on the percentage of bark that they contain. In the manufacturing process, various binding substances and other kinds of additives are used. Particleboards are also affected by the compressing method. The plates can be made by the flat compressing method or by the extruding process. In the first case, the wood chips are largely oriented parallel with the plane of the boards, while in the second case they are oriented perpendicular to the plane. The particleboards may be of monolayer and multilayer structure and consist of layers of chips of various forms and dimensions. The density of particleboards can vary from 0.4 to 1.2 g/cm^3.

As binding materials for particleboards, carbamide and phenol-formaldehyde resins are most commonly used. In addition to the binding agents, such materials as hardeners, water-repellent emulsions, antiseptic agents, and dyes are introduced into the particleboards during the production process. Depending on the type of board, the quantity of binding substances varies from 6 to 13% based on oven-dry wood mass, with averages of about 8–10%. All the above agents affect the dielectric properties of the particleboards.

The dielectric properties of resins vary largely depending on the frequency. For example, the DC of phenol-formaldehyde resin in the dry state and at 25 °C diminishes with increasing frequency. At a frequency of 10^2 Hz, it is 8.2; at $f = 10^6$ Hz, it is 5.4; at $f = 10^8$ Hz, it is 4.4; and at $f = 10^{10}$ Hz, it is 3.52. The loss tangent value of this resin has its maximum (0.135) at a frequency of 10^2 Hz. With increasing frequency, the value of $tg\delta$ decreases, to 0.056 at $f = 10^5$ Hz, and then increases to a maximum value of 0.077 at a frequency of 10^8 Hz. A further increase in frequency causes the $tg\delta$ value to decrease (to 0.037 at a frequency of 10^{10} Hz). Other types of resin may exhibit different responses of ε' and $tg\delta$ to changes in the frequency. For example, urea-formaldehyde resin at 27 °C and with an increase in frequency from 10^2 to 10^{10} Hz shows a decrease in the DC from 6.5 to 4.5; its $tg\delta$ minimum value is equal to 0.020 at $f = 10^5$ Hz, after which it increases to 0.057 at a frequency of 10^{10} Hz. This is the reason why the dielectric parameters of particleboards vary over a wide range depending on their composition, structure, density, moisture content, frequency, and orientation of vector $\bar{E}$ relative to the board plane. Therefore, the approximate measurements are required for a correct determination of the dielectric properties of particleboards in every specific case.

Wood in particleboards makes up more than 86% of the total weight, and therefore the general changes of the dielectric properties of particleboards with density, moisture content, temperature, and frequency are similar in character to those of wood. However, the absolute values of their dielectric properties can differ significantly. A comparison of the DC of particleboards made by flat compressing with that of wood showed that the $\varepsilon'_{\perp}$ of the boards was always lower. This comparison was made at a moisture content $W = 7.5$–20%, a temperature $t = 20\,°C$, and at radiofrequencies, while vector $\bar{E}$ was oriented perpendicular to the board surface. The loss tangent value of particleboards under the same conditions may be higher or lower than the $tg\delta$ value of wood.

Pungs and Lamberts (1954) and Otto (1968) studied the dielectric properties of particleboards in the frequency range from 5×10^6 to 2×10^7 Hz. This choice of frequencies was required by the use of high-frequency heating in the manufacturing process.

Otto (1968) used samples of monolayer particleboards made by the flat compressing process of softwood and hardwood chips. The dimensions of the chips were as follows: length 2–10 mm, width 2–4 mm, and thickness 0.2 mm. The condensation product of urea, formaldehyde, and furfural was used as the resin. The resin quantity in the dry substance was 8–10% of the weight of the dry chips. Ammonium chloride was used as a hardener in a quantity of 1–1.5% of the resin weight. The experiments were carried out at frequencies from 5×10^6 Hz to 2×10^7 Hz. The board density was 0.3–1.0 g/cm^3, the moisture content was 7.5–20%, and vector $\bar{E}$ was oriented perpendicular to the board surface. The dielectric characteristics obtained are given in Table 8.5.

It was established that an increase of the particleboard density from 0.3 to 1.0 g/cm^3 raises the value of $\varepsilon'_{\perp}$ of boards with a moisture content of 7.5–20% by 7–16%, whereas that of $tg\delta_{\perp}$ increases 1.5–3.3 times at $W = 7.5$–15% and by 13–21% at $W = 20\%$.

Figure 8.8 shows the relation between $\varepsilon'_{\perp}$ and $tg\delta_{\perp}$ on the moisture content at a frequency of 5×10^6 Hz and with a density of 0.6 g/cm^3. With increasing moisture content, the increase in the $tg\delta_{\perp}$ is more greater than the increase in the DC.

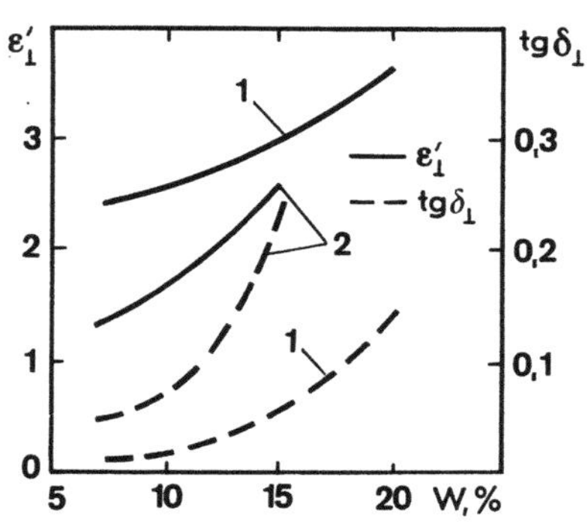

Fig. 8.8. Dependence of $\varepsilon'_{\perp}$ and $tg\delta_{\perp}$ values of particleboards made by the flat compressing process on the moisture content W at $t = 18$–$20\,°C$, density $\rho = 0.6$ g/cm^3, and at frequency $f = 5$ MHz: *1* Otto (1968); *2* Pungs and Lamberts (1954)

Table 8.5. Dielectric parameters $\varepsilon'_\perp$ and $tg\delta_\perp$ of particleboards at frequencies of 5–20 MHz and a temperature of 18 °C. Vector $\bar{E}$ is perpendicular to the plane of the boards (Otto 1968)

W (%)	ρ (g/cm³)	Frequency (MHz)					
		5		15		20	
		$\varepsilon'_\perp$	$tg\delta_\perp$	$\varepsilon'_\perp$	$tg\delta_\perp$	$\varepsilon'_\perp$	$tg\delta_\perp$
7.5	0.3	2.25	0.008	1.95	0.015	1.87	0.018
	0.5	2.35	0.010	2.05	0.018	1.95	0.021
	0.7	2.43	0.012	2.15	0.020	2.05	0.026
	1.0	2.48	0.018	2.20	0.025	2.1	0.030
10	0.3	2.51	0.010	2.24	0.030	2.15	0.035
	0.5	2.60	0.020	2.3	0.034	2.24	0.040
	0.7	2.67	0.025	2.4	0.041	2.3	0.046
	1.0	2.70	0.033	2.5	0.050	2.43	0.055
15	0.3	2.75	0.043	2.55	0.052	2.55	0.064
	0.5	2.85	0.052	2.7	0.068	2.60	0.073
	0.7	2.90	0.060	2.8	0.077	2.70	0.082
	1.0	2.95	0.068	2.87	0.089	2.8	0.095
20	0.3	3.3	0.120	3.05	0.136	2.9	0.142
	0.5	3.53	0.135	3.25	0.147	3.12	0.152
	0.7	3.75	0.138	3.4	0.150	3.25	0.155
	1.0	3.81	0.145	3.47	0.155	3.30	0.160

The difference between the data obtained by Otto (1968) and those reported by Pungs and Lamberts (1954), which is seen in Fig. 8.8, can be attributed to the difference in the wood species used by the investigators as well as to the difference in the structure of the boards and also in the chemical composition of the resins used during the manufacturing process. An increase of moisture content from 7.5 to 20% (at a frequency of 10^7 Hz and at a density of 0.3–1.0 g/cm³) increases the DC from 2.1–2.4 to 3.2–3.7, and the $tg\delta_\perp$ from 0.012–0.020 to 0.127–0.150. An increase of frequency from 5 to 20 MHz at $W = 7.5$–20% results in a decrease of DC by 7–17%, and the $tg\delta_\perp$ value is increased 1.1–2.2 times.

The character of the influence of the temperature of the material on the dielectric characteristics of moist particleboards at a frequency of 17 MHz is shown in Fig. 8.9. An increase of the temperature of the material from 18 to 100 °C at the moisture content 10–20% causes a 1.3-fold increase of the $\varepsilon'_\perp$, while the $tg\delta_\perp$ increases 6 and more times.

In the case of flat compressed boards with the chips placed in layers without ordered orientation, it can be assumed that most of the chips are arranged parallel to the surface and are randomly oriented in the plane of the board. Then, the relations between the dielectric properties of the material with vector

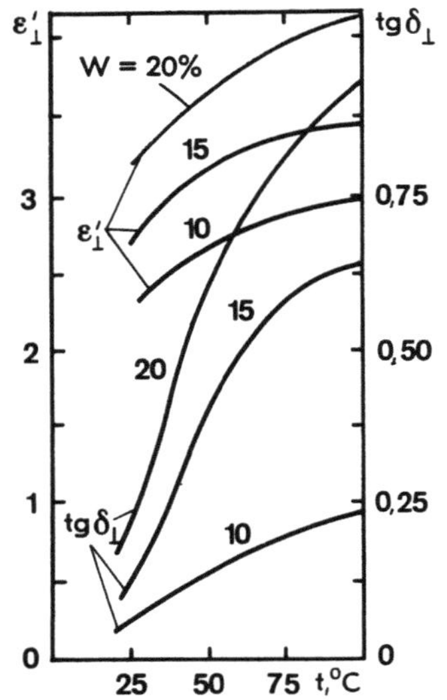

Fig. 8.9. Temperature dependence of $\varepsilon'_\perp$ and $tg\delta_\perp$ values of particleboards made by the flat compressing process with various moisture contents (W) at density $\rho = 0.5$ g/cm^3 and frequency $f = 17$ MHz (Otto 1968)

$\bar{E}$ oriented parallel to the board plane or perpendicular to the plane can accordingly be expressed by Eqs. (8.6) and (8.7) recommended for fiberboards. If the values $\varepsilon'_\perp$ and $tg\delta_\perp$ of the boards are known, using Eqs. (8.6) and (8.7), the values of the DC and loss tangent of the particleboards of flat compression at vector $\bar{E}$ parallel to the sheet plane can be determined.

In the case of ordered orientation of the chips in the plane of the board, the determination of ε' and $tg\delta$ with the vector $\bar{E}$ oriented in the proper direction will require information as to the quantitative allocation factor of the chip distribution in various directions.

In the extrusion type of compression, the chips are oriented mainly perpendicular to the plane of the board, but very likely they are equally oriented in all directions in the transverse plane. Hence, the dielectric parameters ε'_{MD} and $tg\delta_{MD}$ of the extruded particleboard with the vector $\bar{E}$ orientation in the machine direction are the lowest and correspond to the parameters of the flat compressed particleboards with the orientation of vector $\bar{E}$ perpendicular to their surface under otherwise identical conditions. With the orientation of vector $\bar{E}$ in the transverse plane of the extruded particleboard, the values of ε'_{CD} and $tg\delta_{CD}$ are equal in all directions, their values exceeding those in the machine direction.

The relation between the dielectric properties with the orientation of vector $\bar{E}$ in the machine direction (ε'_{MD} and $tg\delta_{MD}$) and in the transverse direction (ε'_{CD} and $tg\delta_{CD}$) can be expressed by formulas similar to Eqs. (8.6) and (8.7)

$$\varepsilon'_{CD} = (1 + k_1)\varepsilon'_{MD}/2 \, , \tag{8.8}$$

$$tg\delta_{CD} = (1 + k_2)tg\delta_{MD}/2 \, , \tag{8.9}$$

where

$k_1 = \varepsilon'_{||}/\varepsilon'_\perp$ for wood (see Tables 4.3, 5.5)

$k_2 = tg\delta_{||}/tg\delta_\perp$ for wood (see Tables 4.3, 5.17).

Table 8.6. Dielectric properties of a material made from wood chips by the extrusion method, using an inorganic binding substance at frequency 2.4 GHz, density $\rho = 1$ g/cm^3 and moisture content $W = 10\%$

t (°C)	Vector $\bar{E}$ oriented in the transverse direction		Vector $\bar{E}$ oriented in the machine direction	
	ε'_{CD}	$tg\,\delta_{CD}$	ε'_{MD}	$tg\,\delta_{MD}$
20	3.07	0.09	2.70	0.07
40	3.13	0.11	2.75	0.09
60	3.24	0.12	2.85	0.10
80	3.39	0.12	2.98	0.10

A determination of the dielectric properties of structural construction beams made of wood chips by the extruding process was carried out at a frequency equal to 2.4 GHz. The following chips were used for manufacture of the material: length 10–25 mm, width 4–10 mm, and thickness up to 0.8 mm. The ratio of components by mass was as follows: wood particles (moisture content 12%) 51.2%, magnesium chloride 14.6%, and magnesium oxide 34.2%. The moisture content of the material was 10%, and the density was 1.0 g/cm^3.

The samples were made in such a way that vector $\bar{E}$, during the measurements in a coaxial waveguide, was oriented in the transverse direction, i.e., so that parameters ε'_{CD} and $tg\delta_{CD}$ could be determined. Average values of the results at temperatures from 20 to 80 °C are given in Table 8.6. The values of the dielectric parameters of the material with vector $\bar{E}$ oriented in the machine direction, ε'_{MD} and $tg\delta_{MD}$, were determined using Eqs. (8.8) and (8.9) with the values of $k_1 = 1.28$ (Table 5.5, $W = 10\%$) and $k_2 = 1.5$ (Table 5.17, $W = 10\%$). When the temperature was increased from 20 to 80 °C, the DC of the material increased by 10%, and the loss tangent value increased by 33–43%.

8.4 Materials Containing Wood, Cellulose, Paper (as a Filler), and Cellulose Derivatives

Wood and wood cellulose are the basic stock for the manufacture of different materials, including plastics. In some materials, wood or cellulose play the part of fillers only and as such on the whole preserve their chemical structure, although in many cases some chemical interactions between the mixed components occur. Other materials are produced by the chemical treatment of wood and cellulose, as a result of which substances with different chemical structures are obtained. The dielectric properties of such materials may differ entirely from those of wood or cellulose.

Figure 8.10 shows the frequency response of the dielectric parameters of three materials: solidified phenol-formaldehyde resin (Bakelite), oven-dry wood with a density of 0.5 g/cm^3, and a composition that comprises resin (46%), wood flour (46%), and various additives (8%).

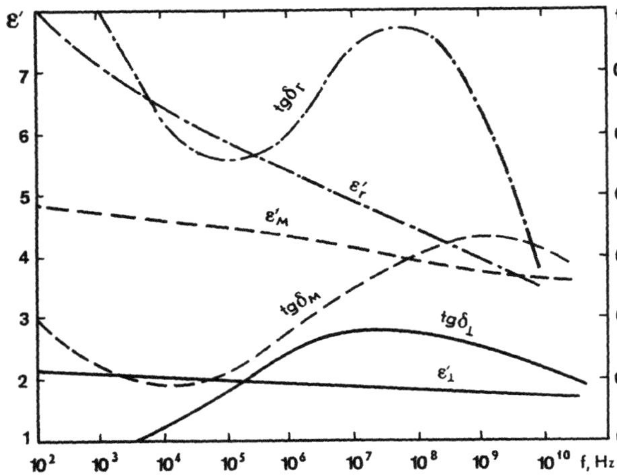

Fig. 8.10. Frequency response of dielectric properties of solidified phenol-formaldehyde resin (values of ε_r' and $tg\,\delta_r$ according to von Hippel 1954), of oven-dry wood with density $\rho_0 = 0.5$ g/cm^3 (curvesw marked $\varepsilon_\perp'$ and $tg\,\delta_\perp$), and of plastic materials (values of ε_M' and $tg\,\delta_M$ according to von Hippel 1954) consisting of phenol-formaldehyde resin (46%), wood flour (46%), and various additives (8%), at $t = 24$–$27\,°$C

At frequencies up to 3×10^9–5×10^9 Hz and at a temperature of 24–27 °C, the values of the DC and the loss tangent of a mixture of equal parts of resin and wood are lower than those of the resin and higher than those of the wood alone. At higher frequencies, the values of the mixture exceed the values of each of the two components. The minimum value of $tg\delta_M$ of the mixture is shifted into the lower frequency range in comparison with the $tg\delta_r$ of the resin (Fig. 8.10), while the maximum point of $tg\delta_M$ of the mixture is shifted into the higher frequency range in comparison with the maxima of the properties of the resin and the wood. With increasing temperature, the picture changes considerably as the parameters of both principal components of the mixture change in their turn as well. Every frequency range possesses its own mechanism of interaction between the electromagnetic field and the substances of the mixture. The interaction takes place on the molecular level, and this mechanism determines the dielectric properties of the material.

Appendix 7 contains the dielectric properties of some plastic materials that include as one component wood, cellulose, or paper. The parameters were measured in the frequency range of 10^2–10^{10} Hz.

Cellulose derivatives possess dielectric properties that differ from those of cellulose itself both in absolute values and in the character of the frequency response of ε' and $tg\delta$. Figure 8.11 shows values of the DC and loss tangent of methyl cellulose, ethylcellulose, and cellulose acetate plotted against frequency in a range from 10^2 to 10^{10} Hz (von Hippel 1954). The curves of the frequency response of the $\varepsilon_{c\perp}'$ and $tg\delta_{c\perp}$ values are also shown in Fig. 8.11. It should be

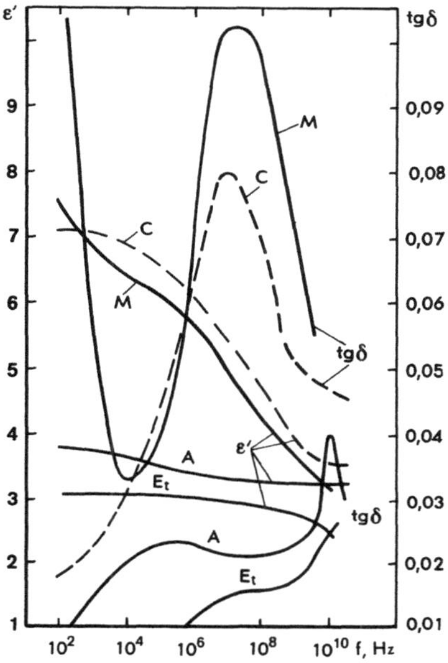

Fig. 8.11. Frequency response of ε' and $tg\delta$ values of cellulose (C), methyl cellulose (M), ethylcellulose (Et), and cellulose acetate (A) at $t = 20-25\,°C$

noted that at frequencies above 10^3 Hz, the DCs of the cellulose derivatives are lower than that of cellulose alone, and that a dispersion of ε' (the decrease in the DC value with an increase of frequency that is characteristic of dielectric materials) takes place. The relationship between the $tg\delta$ values of the cellulose derivatives and the frequency are of a different character (Fig. 8.11) than that of cellulose. This difference is caused by the variation in the chemical composition of the materials.

Chapter 9

Recommendations for the Determination of the Dielectric Properties of Wood and Wood-Based Materials and for Their Use in Calculations

The wide spectrum and intricate character of the correlations between the dielectric properties of wood and wood density, moisture content, temperature, orientation of the vector of the electric field strength relative to the fiber direction, and frequency make it difficult to calculate the interactions between electromagnetic fields and wood. These calculations are required for production processes, designing production facilities, as well as for the development of measuring instruments. In some cases, it is necessary to know the dielectric characteristics at fixed values of moisture content and temperature. During the heating and drying treatment of wood, its temperature or moisture content (or both) change, and usually the average values of the dielectric parameters should be used. Therefore, in order to ensure reliable results and conclusions, it is necessary to determine the correct values of the dielectric properties of a given material.

9.1 Dielectric Properties and the Density of Oven-Dry Wood

The average values of the dielectric characteristics of wood with the electric field strength vector $\overline{E}$ oriented perpendicular to the fiber direction are most closely associated with the initial value of wood density, i.e., with the density of wood in the oven-dry state. The dependence of these characteristics upon tree species is much less pronounced. This is the reason why in Appendices 1 and 3 the dielectric parameters $\varepsilon'_{\perp}$ and $tg\,\delta_{\perp}$ are determined within any frequency range on the basis of the value of the density of oven-dry wood without taking into account the tree species.

The determination of the average values of dielectric characteristics of wood from a given tree species requires knowledge of the wood density in the oven-dry state. The density of the same wood can vary widely as its value depends on the growth conditions, age of the tree, position of the sample in the stem, and other factors. The coefficient of variation of density values of the great majority of wood species averages 10%.

Appendix 8 presents average values of oven-dry wood densities of the industrially most widely used tree species growing in different world regions. These data can be treated as initial values for the determination of the dielectric parameters of various species, employing Appendices 1 and 3 as necessary.

Calculations of the interactions between electromagnetic fields and wood at high frequencies and with microwaves in many cases should not necessarily take into account the difference in the dielectric properties in the radial and tangential directions. It is quite sufficient to know these parameters only in the transverse direction.

However, the influence of the tree species should be taken into consideration if knowledge of the dielectric parameters is required separately in the radial, tangential, or longitudinal directions, or in the case of a specified orientation of vector $\overline{E}$ relative to the principal axes. This influence is taken into account with the aid of empirical coefficients and relations for oven-dry and moist wood presented in Chaps. 4 and 5. The initial data for the determination of ε'_R, ε'_T, ε'_L ($tg\,\delta_R$, $tg\,\delta_T$, $tg\,\delta_L$) values at frequencies of 10^4-10^{10} Hz that will be required for such calculations can be found in Appendices 1 and 3.

9.2 Averaging the Dielectric Parameters Values

A number of measuring instruments operating in radio frequencies and microwave ranges are based on a determination of an attenuation factor in the wood during the passage of electromagnetic waves through the material. Some instruments are based on measurements of a microwave reflection factor. The instruments using these principles are usually intended for measuring the moisture content and thickness of lumber and wood-based materials, detecting defects, checking strength characteristics, estimating surface roughness, as well as inspecting other wood properties. As a rule, the radiation power used for the measurement does not affect the temperature and the moisture content of the material, and therefore in the computation of attenuation and reflection factors, only fixed values of ε' and $tg\,\delta$ are used, these being selected for the appropriate values of temperature, moisture content, and frequency and with a given orientation of the electric field strength vector. The values of these parameters can be found in the tables or calculated with the use of the formulae and diagrams included in this book.

In the process of interactions between the electromagnetic fields and the material that take place under the conditions of variation in temperature, moisture content, or frequency, and with linear (or nearly linear) changes in the dielectric properties under the influence of these factors, the average value (arithmetic mean value) of a given parameter in the given interval of the variation in conditions should be used. For example, on a change in oven-dry wood temperature from 20 to 100 °C, the dependence of the DC of wood on the temperature is linear at any frequency from 10^2 to 10^{10} Hz. Then the average value of the DC at the initial temperature t_0 and at the final temperature t_k is determined as the arithmetic mean value

$$\varepsilon' = (\varepsilon'_{t_0} + \varepsilon'_{t_k})/2\,. \tag{9.1}$$

If the change of the dielectric parameters in the given ranges of temperature, moisture content, or frequency is not linear, the integral values between the given limits should be used for the determination of the dielectric characteristics of wood or of wood-based materials.

During the processes of heating wood in the electromagnetic field that are not accompanied by any changes of moisture content and frequency, the dielectric properties of the wood may change continuously, have maximum and/or minimum values, or change abruptly. For example, with change of the temperature from below to above 0 °C, with a moisture content exceeding the fiber saturation point, and at microwave frequencies, the DC value of wood increases by several times, a phenomenon that is caused by a shift of the free moisture from a solid to a liquid state. In the region of continuous change of a dielectric parameter with the temperature, it is necessary to take the integral over the temperature to find the dielectric constant $(\varepsilon'_t)_i$ and the loss tangent $(tg\,\delta_t)_i$ values using the following equations

$$(\varepsilon'_t)_i = \int_{t_0}^{t_k} \varepsilon'(t)\,dt/(t_k - t_0)\,, \tag{9.2}$$

$$(tg\,\delta_t)_i = \int_{t_0}^{t_k} tg\,\delta(t)\,dt/(t_k - t_0)\,, \tag{9.3}$$

where

t_0 and t_k = initial and final temperatures.

If the parameter has its maximum or minimum value within the heating temperature range, the integral values of the parameter before and after the extreme should be taken. Further calculations should be carried out for the two corresponding parts of the temperature range.

When calculating the defrosting process of wood with subsequent heating in the positive temperature range, the calculations should be carried out separately for temperature ranges below and above 0 °C, with the use of average or integral quantities characterizing the dielectric parameters in the corresponding regions.

The DCs of oven-dry wood have different temperature coefficients in the ranges from − 40 to + 20 °C and from + 20 to + 100 °C, and average values of DC over the temperature range should therefore be taken within these limits. The loss tangent value of oven-dry wood changes curvilinearly with the temperature, with clearly expressed maximum values (Fig. 4.10). Averaging the value of $tg\,\delta$ in accordance with the temperature variations is allowed only in the region of its continuous change in the frequency range from 10^2 to 10^{10} Hz.

At low frequencies up to 10^4 Hz, the dependence of the DC value on temperature is nonlinear over the entire range of moisture content values. This is the reason why in this frequency range only the integral quantities of the DC value should be used. At frequencies of 10^5–10^{10} Hz and at a moisture content of up to 25–30%, the changes of $\varepsilon' = \varphi(t)$ follow dependencies that are nearly linear (Fig. 5.10). Therefore, the value of the DC within the temperature range from − 20 to 90 °C can be accepted as an average between the values of ε' at the initial and the final temperatures. At frequencies of 1–3 GHz, with vector $\overline{E}$ oriented perpendicular to the fiber direction of the wood, the DC value at moisture contents exceeding 30% virtually does not change in the temperature range from 0 to 90 °C. Therefore, in this case, the value of ε' can be considered as constant.

Calculation of the energy released in the material during the drying process at a constant temperature and with a moisture constant changing from an initial value W_0 to a final value W_k should also be carried out with the use of the dielectric parameters $(\varepsilon'_w)_i$ and $(tg\,\delta_w)_i$ determined with the integral taken over the moisture content

$$(\varepsilon'_w)_i = \int_{W_0}^{W_k} \varepsilon'(W)\,dW/(W_k - W_0)\,, \qquad (9.4)$$

$$(tg\,\delta_w)_i = \int_{W_0}^{W_k} tg\,\delta(W)\,dW/(W_k - W_0)\,. \qquad (9.5)$$

At low frequencies from 20 to 10^3 Hz the DC is an exponential function of the moisture content

$$\varepsilon' = a_n W^{b_n}\,, \qquad (9.6)$$

where

a_n and b_n = empirical coefficients given in Table 5.1. The coefficients depend on the tree species and the density and on the orientation of vector $\overline{E}$ and vary for moisture content ranges of 12–24% and 25–100%.

The average DC value is determined as the integral quantity over the moisture content

$$(\varepsilon'_w)_i = [a_n[W_k^{b_n+1} - W_0^{b_n+1})/(b_n + 1)]/(W_k - W_0)\,, \qquad (9.7)$$

where

W_k and W_0 are expressed in percent.

At frequencies above 10^3 Hz and at a moisture content below the fiber saturation point, the dependence of the DC on the moisture content is nonlinear, and in this region the determination of the average value of the DC should be carried out with the integral taken over the moisture content. At a moisture content above the fiber saturation point, the dependence of the DC on the moisture content is almost linear (Fig. 5.5), and in calculations in which the moisture content variations are within 30–120%, the arithmetic mean value of DC in the required range of moisture content changes can be accepted.

Over the entire frequency range from 20 to 10^{10} Hz and in the moisture content region of up to 30%, the dependence $tg\,\delta = \varphi(W)$ is nonlinear, and the integral quantities of the loss tangent value should be used in the calculations. The loss tangent value of wood with a moisture content above 30% and at frequencies above 5×10^7 Hz has a virtually linear dependence on the moisture content (Fig. 5.14), and therefore the arithmetical mean of $tg\,\delta$ within the moisture content region 30–100% can be accepted.

During the process of dielectric drying of wood and wood-based materials, simultaneous changes in temperature and moisture content of the specimen subject to treatment can take place. In order to calculate the energy released in the dielectrics during the drying period, the values of the dielectric properties should be averaged both for the temperature and for the moisture content values within the limits of their changes. It should be noted that averaging the ε' and $tg\,\delta$ values by the two parameters (t, W) is justified if the change of their values in the given temperature and moisture content ranges are continuous.

Determination of the calculated average values of the dielectric properties of compressed and modified wood, paper, and other wood-based materials should be carried out using the formulas for wood depending on the linear or nonlinear character of the given parameter changes within the given range of temperature or moisture content values.

If during the wood treatment the frequency is varied, the wood dielectric parameters should be averaged according to the frequency. Any extension of the ranges of the temperature, moisture content, or frequency averagings results in larger errors in the computations.

Tables and diagrams in this book contain values of dielectric characteristics of wood with the vector of the electric field strength oriented in the direction of the principal anisotropy axes of wood, namely *L*, *R*, and *T*. If the values of the parameters at given angles to principal axes are required, formulas (1.32) and (1.33) can serve this purpose. In the calculations relating to the treatment of such materials as wood chips and shavings and of wood-based materials the fibers of which are not strictly oriented, or in the case of wood processing involving an electromagnetic field with a variable orientation of vector $\overline{E}$, the values of the DC and the loss tangent are to be taken as the average values of $\varepsilon'_{\perp}$ and $\varepsilon'_{\parallel}$ or $tg\,\delta_{\perp}$ and $tg\,\delta_{\parallel}$, respectively.

9.3 Choice of Frequency on Microwave Treatment of Wood

Figure 9.1 shows the electromagnetic spectrum. The upper part shows the wavelength values, and the lower part gives the corresponding frequencies of the electromagnetic radiation. The dielectric properties of wood are most thoroughly studied in the frequency ranges that are used for wood processing and for measuring various characteristics of wood. These lie in the radiofrequency and microwave ranges.

The choice of electromagnetic radiation frequency for the purpose of treating wood and wood-based materials treatment in any specific production process is highly limited, as international agreements allow for the industrial use of only restricted frequency ranges, as is shown in Table 9.1. However, from these ranges, certain frequencies should be chosen to ensure the most effective utilization of the electromagnetic field energy.

For industrial purposes (heating, drying, and glueing of wood) the following frequencies are chiefly used in the most promising microwave range: 0.434, 0.915, 2.45, and 5.8 GHz. The moist wood actively absorbs the electromagnetic field energy at these frequencies. The penetration depth is the factor that

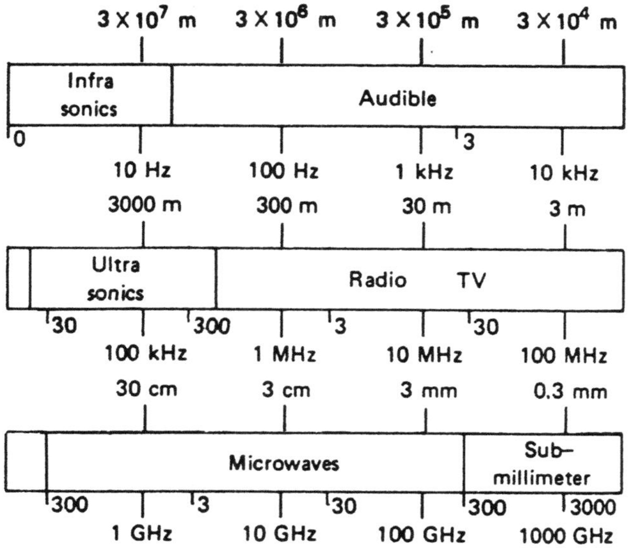

Fig. 9.1. Electromagnetic spectrum

Table 9.1. Frequency allocations for industrial and scientific purposes

Frequency (MHz)	Tolerance (+/−)	Area permitted
0.070	10 KHz	Russia
13.560	0.05%	Worldwide
27.120	0.6%	Worldwide
40.680	0.05%	Worldwide
42, 49, 56, 61, 66	0.2%	UK
84, 168	0.005%	UK
433.92	0.2%	Australia, The Netherlands, Portugal, Germany, Yugoslavia, Switzerland
896	10 MHz	UK
915	13 MHz	USA, Russia
2375	50 MHz	Albania, Bulgaria, Hungary, Romania, Czechoslovakia, Russia
2450	50 MHz	Worldwide except where 2375 MHz is used
3390	0.6%	The Netherlands
5800	75 MHz	Worldwide
6780	0.6%	The Netherlands
24 125	125 MHz	Worldwide
40 680		UK

characterizes the energy absorption and it depends on the dielectric properties of the material. The penetration depth is characterized by the distance d from the surface of the material at which the energy level decreases by $e = 2.72$ times ($1/e = 0.368$) relative to the energy level on the surface.

The value of energy W' at the depth d is to be found from Eq. (1.51)

$$W' = W'_0 \exp(-2\alpha d) , \tag{9.8}$$

where

W'_0 = energy supplied to the surface of the material
α = attenuation constant found through Eq. (1.49).

If $W'/W'_0 = e^{-1}$
then $e^{-1} = e^{-2\alpha d}$ and

$$d = 1/2\alpha . \tag{9.9}$$

The dielectric parameters of wood and the frequency (wavelength λ_0) determine the attenuation constant value

$$\alpha = (2\pi/\lambda_0) [(\varepsilon'/2) (\sqrt{1 + tg^2\delta} - 1)]^{1/2} . \tag{9.10}$$

The results of computation of the penetration depth d according to the wood moisture content at the temperature $t = 20\,°C$ are given in Table 9.2 and are illustrated in Fig. 9.2. With an increase in frequency from 0.434 GHz to 5.8 GHz, depending on the value of moisture content, the penetration depth d decreases 9–18 times.

With a change of the orientation of vector $\bar{E}$ from perpendicular to parallel relative to the direction of the softwood fibers, the penetration depth decreases

Table 9.2. Penetration depth d at which incident power drops to $1/e$ (about 37%) ($\rho_0 = 0.5$ g/cm^3, $t = 20\,°C$)

W (%)	Vector $\bar{E}$ orientation	d (cm) at frequency (GHz) 0.434	0.915	2.45	5.8
Softwood and hardwood					
	⊥ to grain				
0		305	156	66.7	28.9
10		66.7	33.3	13.9	5.7
20		40.7	17.9	6.5	2.5
30		26.9	11.1	4.7	1.6
60		19.4	9.4	3.9	1.4
100		14.9	8.5	2.9	1.0
Softwood					
	‖ to grain				
0		171	94.3	42.7	19.1
10		42	22.3	7.2	3.1
20		25	11.7	3.9	1.4
30		15.9	6.9	2.8	0.9
60		11.6	5.9	2.6	0.8
100		9.1	5.3	1.9	0.6

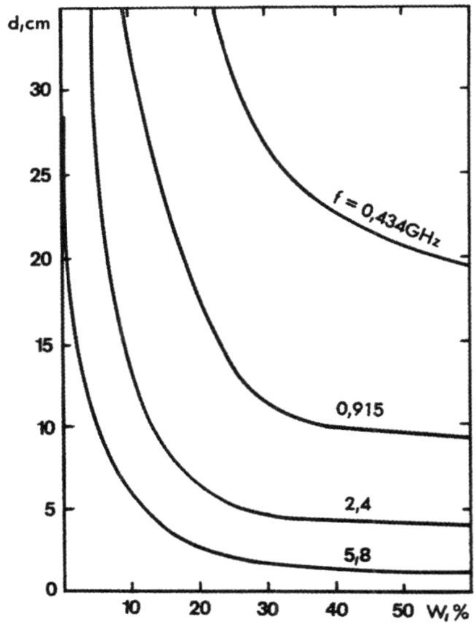

Fig. 9.2. Dependence of penetration depth d on the wood moisture content (W, %), with wood density $\rho_0 = 0.5$ g/cm^3 at microwave frequencies (f, GHz) when vector $\overline{E}$ is oriented perpendicular to the fiber direction and at the temperature $t = 20$ °C

1.5–1.9 times (at $t = 20$ °C). At frequencies of 0.915 GHz and higher, the most important decrease of the penetration depth occurs at moisture contents up to 30%, after which the change of the d value is very small. This can be explained by a decrease of the loss tangent value at wood moisture contents exceeding the fiber saturation point.

The increase of temperature from 20 to 90 °C at a frequency of 0.434 GHz causes a decrease in the penetration depth in the moist wood (Fig. 9.3), which is connected with an increase in the $tg\delta$ with increasing temperature at this frequency. At frequencies of 0.915, 2.45, and 5.8 GHz under the same conditions,

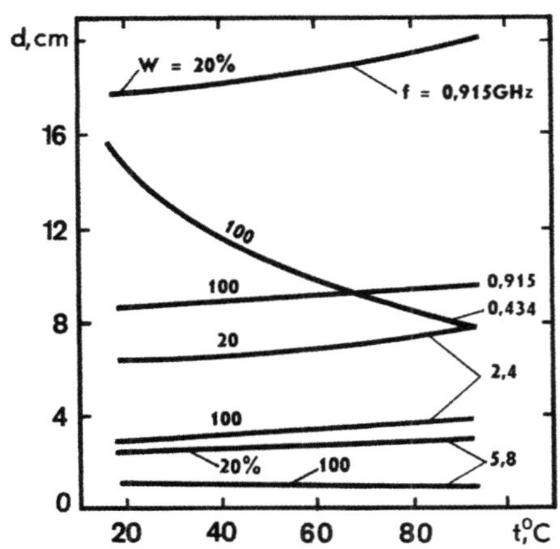

Fig. 9.3. Dependence of penetration depth d on the temperature (t, °C) of wood with moisture content $W = 20$ and 100%, density $\rho_0 = 0.5$ g/cm^3, and vector $\overline{E}$ oriented perpendicular to the fiber direction at microwave frequencies (f)

the penetration depth increases slightly, which can be explained by a decrease in $tg\,\delta$ with increasing temperature. At negative temperatures, the dielectric properties of wood at frequencies of 0.915–5.8 GHz decrease, which results in an increase in the penetration depth value in comparison with its value at positive temperatures. For example, the penetration depth for wood with a density of 0.5 g/cm^3 and a moisture content of 10–100%, at $t = -20\,°C$ is higher than that at $t = +20\,°C$; at a frequency of 0.915 GHz it is 1.1–1.5 times higher, and at a frequency of 2.45 GHz it is 1.3–2.0 times higher, and at a frequency of 5.8 GHz it is 1.6–2.8 times higher.

Analysis of the dependence of the penetration depth on frequency, wood moisture content, and density, as well as on vector $\overline{E}$ orientation relative to the fiber direction, ensures the proper choice of frequency for the most expedient use in any specific process of wood treatment. For heating thick lumber (rough logs, beams, packaged boards) throughout their depth and for preserving the uniform supply of energy, the frequency 0.434 GHz is preferable, while for sawn timber, fiberboards, and particle boards, frequencies of 0.915 and 2.45 GHz are more suitable, and for veneer, paper, and board, frequencies of 2.45 and 5.8 GHz are mostly used.

Appendix 1

Dielectric properties of oven-dry wood and cell wall substance (electric field $\bar{E}$ is across the grain, $t = 20$–$25\,^{\circ}$C)

f (Hz)	0	10^2			10^3			10^4			10^5		
ρ_0 (g/cm^3)	$\varepsilon'_{\perp S}$	$\varepsilon'_\perp$	$tg\delta_\perp$	$\varepsilon''_\perp$	$\varepsilon'_\perp$	$tg\delta_\perp$	$\varepsilon''_\perp$	$\varepsilon'_\perp$	$tg\delta_\perp$	$\varepsilon''_\perp$	$\varepsilon'_\perp$	$tg\delta_\perp$	$\varepsilon''_\perp$
0.13	1.4	1.4	0.005	0.007	1.4	0.005	0.007	1.4	0.005	0.007	1.4	0.007	0.010
0.2	1.55	1.5	0.005	0.008	1.5	0.005	0.008	1.5	0.006	0.009	1.5	0.010	0.015
0.3	1.7	1.7	0.005	0.009	1.7	0.006	0.010	1.7	0.008	0.014	1.6	0.012	0.019
0.4	2.0	1.9	0.006	0.011	1.9	0.007	0.013	1.9	0.010	0.019	1.8	0.016	0.029
0.5	2.3	2.2	0.006	0.013	2.1	0.008	0.017	2.1	0.012	0.025	2.1	0.018	0.038
0.6	2.5	2.5	0.007	0.018	2.4	0.010	0.024	2.4	0.014	0.034	2.3	0.021	0.048
0.7	2.9	2.8	0.007	0.020	2.7	0.011	0.030	2.7	0.016	0.043	2.6	0.023	0.060
0.8	3.2	3.1	0.008	0.025	3.1	0.012	0.037	3.0	0.017	0.051	2.9	0.026	0.075
0.9	3.6	3.5	0.008	0.028	3.4	0.013	0.044	3.3	0.019	0.063	3.2	0.028	0.090
1.0	4.0	3.9	0.009	0.035	3.8	0.014	0.053	3.8	0.020	0.076	3.6	0.029	0.104
1.1	4.5	4.4	0.010	0.044	4.3	0.014	0.060	4.2	0.021	0.088	4.0	0.031	0.124
1.2	4.9	4.8	0.011	0.053	4.7	0.015	0.071	4.6	0.022	0.101	4.3	0.033	0.142
1.3	5.5	5.3	0.011	0.058	5.2	0.016	0.083	5.1	0.023	0.117	4.8	0.034	0.163
1.4	6.0	5.8	0.012	0.070	5.7	0.017	0.097	5.5	0.024	0.132	5.2	0.035	0.182
1.53 (C.W.S)	6.8	6.5	0.014	0.091	6.4	0.018	0.115	6.3	0.025	0.158	5.9	0.037	0.218

f (Hz)	10^6			5.10^6			10^7			10^8		
ρ_0 (g/cm^3)	$\varepsilon'_\perp$	$tg\delta_\perp$	$\varepsilon''_\perp$	$\varepsilon'_\perp$	$tg\delta_\perp$	$\varepsilon''_\perp$	$\varepsilon'_\perp$	$tg\delta_\perp$	$\varepsilon''_\perp$	$\varepsilon'_\perp$	$tg\delta_\perp$	$\varepsilon''_\perp$
0.13	1.4	0.013	0.018	1.3	0.015	0.020	1.3	0.016	0.021	1.3	0.014	0.018
0.2	1.4	0.016	0.022	1.4	0.017	0.024	1.4	0.018	0.025	1.4	0.017	0.024
0.3	1.5	0.019	0.029	1.5	0.021	0.032	1.5	0.021	0.032	1.5	0.020	0.030
0.4	1.7	0.022	0.037	1.7	0.024	0.041	1.7	0.025	0.043	1.7	0.024	0.041
0.5	2.0	0.025	0.050	1.9	0.028	0.053	1.9	0.028	0.053	1.9	0.028	0.053
0.6	2.2	0.028	0.062	2.2	0.031	0.068	2.2	0.031	0.068	2.1	0.031	0.065
0.7	2.4	0.031	0.074	2.4	0.035	0.084	2.4	0.035	0.084	2.3	0.034	0.078
0.8	2.7	0.034	0.092	2.6	0.038	0.099	2.6	0.038	0.099	2.5	0.038	0.095
0.9	3.0	0.037	0.111	3.0	0.041	0.123	2.9	0.042	0.122	2.7	0.041	0.111
1.0	3.3	0.040	0.132	3.3	0.044	0.145	3.2	0.046	0.147	3.0	0.044	0.132
1.1	3.6	0.042	0.151	3.6	0.049	0.176	3.5	0.050	0.175	3.3	0.047	0.155
1.2	4.0	0.045	0.180	3.9	0.052	0.203	3.8	0.054	0.205	3.5	0.051	0.179
1.3	4.4	0.048	0.211	4.3	0.056	0.241	4.2	0.058	0.244	3.8	0.054	0.205
1.4	4.9	0.050	0.245	4.7	0.059	0.277	4.5	0.062	0.279	4.1	0.057	0.234
1.53 (C.W.S)	5.5	0.053	0.292	5.2	0.063	0.328	5.0	0.067	0.335	4.5	0.061	0.275

f (Hz)	10^9			3.10^9			10^{10}			10^{11}			∞
ρ_0 (g/cm^3)	$\varepsilon'_\perp$	$tg\delta_\perp$	$\varepsilon''_\perp$	$\varepsilon'_\perp$	$tg\delta_\perp$	$\varepsilon''_\perp$	$\varepsilon'_\perp$	$tg\delta_\perp$	$\varepsilon''_\perp$	$\varepsilon'_\perp$	$tg\delta_\perp$	$\varepsilon''_\perp$	$\varepsilon'_{\perp\infty}$
0.13	1.2	0.009	0.011	1.2	0.007	0.008	1.2	0.007	0.008	1.2	0.006	0.007	1.16
0.2	1.3	0.013	0.017	1.3	0.011	0.014	1.3	0.010	0.013	1.2	0.008	0.010	1.2
0.3	1.5	0.017	0.026	1.4	0.015	0.021	1.4	0.014	0.020	1.4	0.012	0.017	1.3
0.4	1.7	0.022	0.037	1.6	0.019	0.030	1.5	0.018	0.027	1.5	0.015	0.023	1.4
0.5	1.8	0.025	0.045	1.7	0.023	0.039	1.7	0.022	0.037	1.6	0.018	0.029	1.5
0.6	2.0	0.029	0.058	1.9	0.027	0.051	1.8	0.026	0.047	1.8	0.022	0.040	1.65
0.7	2.2	0.032	0.070	2.1	0.030	0.063	2.0	0.029	0.058	2.0	0.025	0.050	1.8
0.8	2.3	0.036	0.083	2.2	0.033	0.073	2.2	0.032	0.070	2.2	0.028	0.062	2.0
0.9	2.5	0.038	0.095	2.4	0.035	0.084	2.4	0.033	0.079	2.4	0.030	0.072	2.1
1.0	2.7	0.039	0.105	2.5	0.037	0.093	2.5	0.036	0.090	2.5	0.032	0.080	2.3
1.1	2.9	0.041	0.119	2.7	0.039	0.105	2.7	0.039	0.105	2.6	0.034	0.088	2.45
1.2	3.0	0.044	0.132	2.9	0.041	0.119	2.8	0.040	0.112	2.7	0.035	0.095	2.5
1.3	3.1	0.046	0.143	3.0	0.043	0.129	2.9	0.041	0.119	2.8	0.036	0.101	2.6
1.4	3.3	0.047	0.155	3.1	0.044	0.136	3.1	0.042	0.130	3.0	0.038	0.114	2.8
1.53 (C.W.S.)	3.5	0.049	0.172	3.3	0.046	0.152	3.3	0.043	0.142	3.1	0.039	0.121	2.9

Appendix 2

Dielectric constant (ε') and loss tangent ($tg\delta$) of wood at frequencies of 20–1000 Hz (James 1975)

Moisture content (%)				0		10		20		30		60		100	
f (Hz)	ρ_0 (g/cm^3)	Orientation $\bar{E}$	t (°C)	ε'	$tg\delta$	ε'	$tg\delta$	ε'	$tg\delta$	ε'	$tg\delta$	ε'	$tg\delta$	ε'	$tg\delta$
Coniferous wood (*Douglas-fir*)															
20	0.4–0.5	*R*	−20	2.6	0.063	6.6	0.8	44	27	380	45	1680	32	3400	14.5
			+25	2.8	0.040	15	13.5	2100	35	5200	38	12400	41	22000	44
			45	2.8	0.23	34	22	5700	31	19000	29	58000	29	111000	28
			90	5.6	0.80	205	47	29000	36	95000	22	310000	21	600000	19.5
10^2	0.4–0.5	*R*	−20	2.6	0.038	5.5	0.4	23	15	140	29.5	640	21	1300	10
			+25	2.7	0.025	9.6	5	275	47	660	61	1540	66	2700	72
			45	2.7	0.089	13	13.5	710	45	2200	51	6600	50	12500	49
			90	3.7	0.45	31	51	3400	56	12000	42	38000	38	73000	33
10^3	0.4–0.5	*R*	−20	2.4	0.060	4.4	0.15	12.8	3.2	29	5.8	80	5.0	150	3.9
			+25	2.4	0.021	5.8	1.2	30	40	57	69	144	72	260	76
			45	2.5	0.021	7.6	2.5	56	49	130	75	360	80	670	89
			90	2.9	0.10	10	14.5	184	72	580	82	1840	76	3560	68

White oak																
20	0.65–0.75	*R*	−20	2.9	0.041	8.4	0.86	120	20	450	25	1690	17	3340	7.2
			+25	3.0	0.037	28	16	1340	35	8100	33	21000	29	42000	24
			45	3.1	0.038	81	22	7000	32	27000	34	103000	28	204000	19
			90	3.8	0.12	680	39	30500	37	120000	23	460000	19.5	910000	14.5
10^2	0.65–0.75	*R*	−20	2.6	0.016	7.1	0.4	50	9.6	160	15	550	11	1070	4.8
			+25	2.8	0.014	8	6.5	320	41	1400	42	3500	36	7000	29
			45	2.8	0.032	26	14.5	1100	44	3500	47	12200	39	23500	30
			90	3.4	0.054	73	51	3300	63	13000	42	53000	35	106000	25
10^3	0.65–0.75	*R*	−20	2.4	0.044	5.7	0.19	20	2.9	40	5.1	100	4.2	170	2.8
			+25	2.5	0.021	6	1.4	80	23	250	24	820	19	1900	13
			45	2.6	0.019	12	4.4	130	37	560	37	2350	29	4700	18
			90	3.1	0.034	13	23	200	89	700	70	3200	56	6400	39

Appendix 3

Dielectric constant ($\varepsilon'_\perp$) and loss tangent ($tg\delta_\perp$) of wood at frequencies of 10^4–10^{10} Hz (vector $\bar{E}$ orientation is perpendicular to the grain, moisture content 0–100 %, temperature −20 to +90 °C)

f (Hz)	Density in oven-dry condition	t (°C)	Moisture content (%)											
			0		10		20		30		60		100	
	ρ_0 (g/cm³)		$\varepsilon'_\perp$	$tg\delta_\perp$	$\varepsilon'_\perp$	$tg\delta_\perp$	$\varepsilon'_\perp$	$tg\delta_\perp$	$\varepsilon'_\perp$	$tg\delta_\perp$	$\varepsilon'_\perp$	$tg\delta_\perp$	$\varepsilon'_\perp$	$tg\delta_\perp$
10^4	0.3	−20	1.6	0.013	2.4	0.04	4.9	0.3	6.0	0.6	8.5	0.75	10.6	1.05
		+20	1.7	0.008	2.9	0.09	7.7	3.3	13.7	6	39	6	56	6
		50	1.8	0.007	3.2	0.31	9.8	8.3	19.5	13.2	59	11.4	90	9.6
		90	2.0	0.005	3.7	0.59	12.6	14.9	–	22.8	85	18.6	134	14.4
	0.4	−20	1.8	0.016	3.0	0.06	6.3	0.4	7.9	0.8	11	1.0	13.8	1.4
		+20	1.9	0.010	3.6	0.12	9.9	4.4	18	8	52	8	74	8
		50	2.0	0.008	4.0	0.41	12.6	11.0	26	17.6	79	15.2	118	12.8
		90	2.2	0.006	4.6	0.79	16.1	19.8	–	30.4	114	24.8	178	19.2
	0.5	−20	2.0	0.019	3.5	0.07	7.7	0.5	9.7	1.0	13.5	1.3	17	1.8
		+20	2.1	0.012	4.2	0.15	12.1	5.5	22	10	65	10	92	10
		50	2.3	0.010	4.7	0.51	15.4	13.8	31	22	98	19	147	16
		90	2.5	0.007	5.4	0.99	19.7	24.8	–	38	142	31	221	24
	0.6	−20	2.3	0.022	4.0	0.08	9.2	0.6	11.4	1.2	16	1.5	20.2	2.1
		+20	2.4	0.014	4.8	0.18	14.4	6.6	26	12	77	12	109	12
		50	2.6	0.011	5.4	0.61	18.3	16.5	37	26.4	116	22.8	174	19.2
		90	2.8	0.008	6.1	1.19	23.5	29.7	–	45.6	168	37.2	262	29
	0.7	−20	2.5	0.026	4.6	0.10	10.9	0.7	13.6	1.4	18.5	1.75	23.4	2.5
		+20	2.7	0.016	5.5	0.21	17	7.7	31	14	90	14	123	14
		50	2.9	0.013	6.2	0.71	22	19.25	44	30.8	136	26.6	197	22.4
		90	3.2	0.009	7.0	1.39	28	34.7	–	53.2	197	43.4	295	33.6

	0.8	−20	2.8	0.027	5.1	0.11	12.2	0.8	15.4	1.6	21	2.0	27	2.8
		+20	3.0	0.017	6.1	0.24	19	8.8	35	16	103	16	137	16
		50	3.2	0.014	6.8	0.82	24.1	22	49.7	35.2	156	30.4	219	25.6
		90	3.5	0.010	7.8	1.6	31	39.6	–	60.8	226	49.6	329	38.4
10^5	0.3	−20	1.5	0.017	2.5	0.03	4.2	0.13	4.7	0.21	5.3	0.27	6.1	0.36
		+20	1.6	0.012	2.8	0.03	5.3	0.57	10.4	0.9	23	1.2	35	1.2
		50	1.7	0.009	3.1	0.08	6.1	1.51	12.0	2.3	27	2.3	43	2.1
		90	1.9	0.005	3.4	0.16	7.2	2.8	14.0	4.1	33	3.7	55	3.3
	0.4	−20	1.7	0.022	3.0	0.04	5.4	0.18	5.7	0.28	6.7	0.36	7.8	0.48
		+20	1.8	0.016	3.4	0.04	6.7	0.76	13.6	1.2	31	1.6	48	1.6
		50	1.9	0.012	3.7	0.11	7.7	2.01	15.6	3.0	37	3.0	60	2.8
		90	2.1	0.007	4.1	0.21	9.1	3.7	18.4	5.4	44	5.0	75	4.4
	0.5	−20	2.0	0.025	3.5	0.06	6.6	0.22	7.0	0.35	8.1	0.45	9.5	0.60
		+20	2.1	0.018	4.0	0.05	8.2	0.95	17	1.5	38	2.1	58	2.1
		50	2.3	0.014	4.4	0.14	9.4	2.5	19.6	3.8	45	4.0	72	3.7
		90	2.5	0.008	4.8	0.27	11.1	4.6	23	6.8	54	6.5	90	5.8
	0.6	−20	2.2	0.029	4.0	0.07	7.7	0.26	8.1	0.42	9.5	0.54	11.2	0.72
		+20	2.3	0.021	4.5	0.05	9.6	1.14	20	1.8	46	2.5	69	2.5
		50	2.5	0.016	4.9	0.17	11	3.0	23	4.5	54	4.8	86	4.4
		90	2.7	0.009	5.4	0.32	13	5.5	27	8.1	65	7.8	108	6.9
	0.7	−20	2.4	0.032	4.5	0.08	8.8	0.31	9.3	0.49	11	0.63	13	0.84
		+20	2.6	0.023	5.1	0.06	11	1.33	23	2.1	53	2.9	80	2.9
		50	2.8	0.017	5.6	0.19	12.7	3.5	26	5.3	63	5.5	99	5.1
		90	3.1	0.010	6.2	0.37	15	6.5	31	9.5	75	9.0	125	8.0
	0.8	−20	2.7	0.036	5.0	0.09	10	0.35	10.8	0.56	11.4	0.72	14.6	0.96
		+20	2.9	0.026	5.7	0.07	12.5	1.52	26	2.4	61	3.3	85	–
		50	3.1	0.020	6.2	0.22	14.4	4.0	29.9	6.0	72	6.3	105.4	–
		90	3.4	0.011	6.9	0.42	16.9	7.4	35.1	10.8	86.6	10.2	132.6	–

Appendix 3 (Continued)

f (Hz)	Density in oven-dry condition ρ_0 (g/cm³)	t (°C)	Moisture content (%) 0 $\varepsilon'_\perp$	0 $tg\delta_\perp$	10 $\varepsilon'_\perp$	10 $tg\delta_\perp$	20 $\varepsilon'_\perp$	20 $tg\delta_\perp$	30 $\varepsilon'_\perp$	30 $tg\delta_\perp$	60 $\varepsilon'_\perp$	60 $tg\delta_\perp$	100 $\varepsilon'_\perp$	100 $tg\delta_\perp$
10^6	0.3	−20	1.4	0.020	2.1	0.04	2.7	0.07	3.4	0.09	4.4	0.09	5.1	0.08
		+20	1.5	0.019	2.3	0.03	3.4	0.14	5.5	0.28	11.5	0.4	20	0.4
		50	1.6	0.015	2.4	0.05	3.9	0.27	6.3	0.53	13.2	0.77	23	0.8
		90	1.8	0.010	2.6	0.08	4.6	0.43	7.4	0.87	15.5	1.27	27	1.32
	0.4	−20	1.6	0.023	2.5	0.05	3.4	0.10	4.0	0.12	5.5	0.12	6.5	0.11
		+20	1.7	0.022	2.7	0.04	4.2	0.19	7	0.37	15	0.54	26	0.54
		50	1.8	0.017	2.9	0.07	4.8	0.36	8.1	0.70	17	1.04	30	1.07
		90	2.0	0.011	3.1	0.11	5.7	0.59	9.5	1.15	20	1.7	35	1.8
	0.5	−20	1.9	0.026	2.9	0.07	4.1	0.12	4.7	0.15	6.7	0.15	7.9	0.14
		+20	2.0	0.025	3.1	0.05	5.1	0.24	8.5	0.46	19	0.67	32	0.67
		50	2.2	0.020	3.3	0.09	5.9	0.46	9.8	0.87	22	1.29	37	1.33
		90	2.4	0.013	3.5	0.14	6.9	0.74	11.5	1.43	26	2.1	43	2.2
	0.6	−20	2.1	0.029	3.2	0.08	4.7	0.14	5.6	0.18	7.8	0.18	9.3	0.16
		+20	2.2	0.028	3.5	0.05	5.9	0.29	10	0.56	22	0.80	38	0.80
		50	2.4	0.022	3.7	0.10	6.8	0.55	11.5	1.06	25	1.5	44	1.6
		90	2.6	0.014	4.0	0.17	8.0	0.90	13.5	1.7	30	2.5	51	2.6
	0.7	−20	2.3	0.032	3.6	0.09	5.4	0.17	6.2	0.21	8.9	0.21	10.7	0.19
		+20	2.4	0.031	3.9	0.06	6.7	0.34	11.5	0.64	26	0.94	44	0.94
		50	2.6	0.024	4.1	0.12	7.7	0.65	13.2	1.22	30	1.8	51	1.9
		90	2.8	0.016	4.5	0.19	9.0	1.05	15.5	1.9	35	3.0	59	3.1
	0.8	−20	2.5	0.035	4.0	0.10	6.0	0.19	8.5	0.24	10	0.24	12	0.22
		+20	2.7	0.034	4.4	0.07	7.5	0.38	13	0.74	29	1.08	46	–
		50	2.9	0.027	4.7	0.14	8.6	0.72	15	1.41	33.4	2.08	53	–
		90	3.2	0.017	5.0	0.22	10.1	1.18	17.6	2.30	39	3.42	62	–

10^7	0.3	−20	1.4	0.017	2.1	0.06	2.9	0.06	3.4	0.06	3.8	0.05	4.5	0.05
		+20	1.5	0.021	2.2	0.03	3.2	0.07	4.4	0.10	10.4	0.10	18	0.10
		50	1.6	0.020	2.3	0.04	3.4	0.11	4.7	0.17	11.0	0.19	18.5	0.21
		90	1.7	0.018	2.35	0.05	3.6	0.15	5.0	0.27	11.9	0.31	19.3	0.36
	0.4	−20	1.6	0.020	2.5	0.08	3.4	0.08	4.2	0.08	4.7	0.07	5.7	0.06
		+20	1.7	0.025	2.6	0.04	4.0	0.09	5.6	0.14	13.7	0.14	24	0.14
		50	1.8	0.024	2.7	0.05	4.2	0.14	5.9	0.24	14.5	0.27	25	0.30
		90	1.9	0.021	2.8	0.07	4.6	0.20	6.4	0.38	15.6	0.43	26	0.50
	0.5	−20	1.8	0.022	2.9	0.11	4.1	0.10	5.1	0.10	5.7	0.09	6.9	0.08
		+20	1.9	0.028	3.0	0.05	4.7	0.11	6.7	0.17	17	0.17	30	0.17
		50	2.0	0.026	3.1	0.07	5.0	0.17	7.1	0.29	18	0.32	31	0.36
		90	2.1	0.024	3.2	0.09	5.4	0.24	7.6	0.46	19	0.53	32	0.61
	0.6	−20	2.1	0.025	3.3	0.13	4.7	0.12	5.9	0.12	6.6	0.11	8.1	0.09
		+20	2.2	0.031	3.4	0.06	5.4	0.13	7.8	0.20	20	0.20	36	0.20
		50	2.3	0.029	3.5	0.08	5.7	0.20	8.3	0.34	21	0.38	37	0.42
		90	2.4	0.027	3.6	0.10	6.2	0.28	8.9	0.54	23	0.62	39	0.72
	0.7	−20	2.3	0.028	3.6	0.15	5.4	0.14	6.7	0.14	7.5	0.14	9.3	0.10
		+20	2.4	0.035	3.8	0.07	6.2	0.15	9.0	0.24	23	0.24	42	0.24
		50	2.5	0.033	3.9	0.09	6.6	0.23	9.5	0.41	24	0.46	43	0.51
		90	2.7	0.030	4.1	0.12	7.1	0.33	10.3	0.64	26	0.74	45	0.86
	0.8	−20	2.5	0.030	4.0	0.17	6.3	0.16	7.5	0.16	8.4	0.14	10.4	0.12
		+20	2.6	0.038	4.2	0.08	6.9	0.18	10.1	0.27	26	0.27	43	0.27
		50	2.7	0.036	4.3	0.10	7.3	0.27	10.7	0.46	27.6	0.51	44	0.57
		90	2.9	0.033	4.5	0.14	7.9	0.39	11.5	0.72	29.6	0.84	46	0.97
10^8	0.3	−20	1.4	0.014	2.0	0.08	2.7	0.12	2.9	0.12	3.4	0.12	40	0.10
		+20	1.5	0.020	2.1	0.06	2.9	0.07	3.6	0.09	6.4	0.11	11.7	0.12
		50	1.6	0.022	2.2	0.05	3.1	0.10	3.8	0.15	6.8	0.19	12.1	0.21
		90	1.7	0.026	2.25	0.04	3.3	0.14	4.1	0.23	7.3	0.29	12.5	0.33

Appendix 3 (Continued)

f (Hz)	Density in oven-dry condition ρ_0 (g/cm^3)	t (°C)	Moisture content (%) 0 $\varepsilon'_\perp$	0 $tg\delta_\perp$	10 $\varepsilon'_\perp$	10 $tg\delta_\perp$	20 $\varepsilon'_\perp$	20 $tg\delta_\perp$	30 $\varepsilon'_\perp$	30 $tg\delta_\perp$	60 $\varepsilon'_\perp$	60 $tg\delta_\perp$	100 $\varepsilon'_\perp$	100 $tg\delta_\perp$
	0.4	−20	1.6	0.017	2.4	0.10	3.2	0.16	3.6	0.16	4.2	0.16	5.0	0.13
		+20	1.7	0.024	2.5	0.08	3.5	0.10	4.3	0.12	7.4	0.14	16	0.16
		50	1.8	0.027	2.6	0.07	3.7	0.15	4.6	0.20	7.8	0.24	16.5	0.28
		90	1.9	0.031	2.7	0.05	4.0	0.21	4.9	0.31	8.4	0.38	17.1	0.44
	0.5	−20	1.8	0.020	2.8	0.13	3.8	0.20	4.2	0.20	5.0	0.20	6.0	0.17
		+20	1.9	0.028	2.9	0.10	4.1	0.12	5.1	0.15	10.1	0.18	19	0.20
		50	2.0	0.031	3.0	0.09	4.3	0.17	5.4	0.25	10.7	0.31	19.6	0.35
		90	2.1	0.036	3.1	0.07	4.7	0.25	5.8	0.39	11.5	0.48	20	0.55
	0.6	−20	2.0	0.022	3.1	0.16	4.3	0.24	4.8	0.24	5.7	0.24	6.9	0.20
		+20	2.1	0.031	3.2	0.12	4.7	0.14	5.9	0.17	11.9	0.22	23	0.24
		50	2.2	0.035	3.3	0.10	5.0	0.20	6.3	0.29	12.6	0.38	24	0.42
		90	2.3	0.040	3.4	0.08	5.4	0.29	6.7	0.44	13.6	0.59	25	0.66
	0.7	−20	2.2	0.025	3.5	0.18	4.9	0.28	5.5	0.28	6.5	0.28	7.9	0.23
		+20	2.3	0.034	3.6	0.14	5.3	0.17	6.7	0.20	13.9	0.25	26	0.28
		50	2.4	0.038	3.7	0.12	5.6	0.25	7.1	0.34	14.7	0.43	27	0.49
		90	2.5	0.044	3.9	0.09	6.0	0.35	7.6	0.52	15.8	0.67	30	0.77
	0.8	−20	2.4	0.027	3.8	0.21	5.5	0.32	6.1	0.32	7.3	0.32	8.9	0.26
		+20	2.5	0.038	4.0	0.16	6.0	0.19	7.6	0.23	16	0.29	27	0.32
		50	2.6	0.043	4.1	0.14	6.4	0.28	8.1	0.39	17.0	0.50	28	0.56
		90	2.8	0.049	4.3	0.10	6.8	0.39	8.7	0.60	18.2	0.78	29	0.88
10^9	0.3	−20	1.4	0.012	1.5	0.05	1.6	0.08	1.7	0.12	2.0	0.12	2.3	0.12
		+20	1.5	0.017	1.9	0.06	2.4	0.09	3.2	0.13	4.3	0.12	8.0	0.10
		50	1.6	0.019	2.0	0.06	2.44	0.08	3.2	0.12	4.3	0.11	8.0	0.10
		90	1.7	0.022	2.2	0.06	2.5	0.08	3.2	0.11	4.3	0.10	8.0	0.09

	0.4	−20	1.6	0.016	1.7	0.06	1.8	0.11	1.9	0.16	2.3	0.16	2.7	0.16
		+20	1.7	0.022	2.2	0.08	2.9	0.12	3.9	0.17	5.6	0.16	10.5	0.13
		50	1.8	0.025	2.3	0.08	2.9	0.11	3.9	0.16	5.6	0.15	10.5	0.12
		90	1.9	0.028	2.5	0.08	3.0	0.10	3.9	0.14	5.6	0.13	10.5	0.11
	0.5	−20	1.7	0.018	1.9	0.08	2.0	0.14	2.2	0.20	2.6	0.20	3.1	0.20
		+20	1.8	0.025	2.5	0.10	3.4	0.16	4.7	0.22	6.9	0.21	13.1	0.17
		50	1.9	0.028	2.7	0.10	3.45	0.15	4.7	0.20	6.9	0.20	13.1	0.16
		90	2.0	0.032	2.9	0.10	3.5	0.14	4.7	0.18	6.9	0.18	13.1	0.15
	0.6	−20	1.9	0.021	2.0	0.09	2.2	0.17	2.4	0.24	2.9	0.24	3.5	0.24
		+20	2.0	0.029	2.7	0.12	3.9	0.19	5.4	0.26	8.1	0.24	15.6	0.20
		50	2.1	0.032	2.9	0.12	4.0	0.18	5.4	0.24	8.1	0.22	15.6	0.19
		90	2.2	0.037	3.1	0.12	4.0	0.16	5.4	0.21	8.1	0.20	15.6	0.18
	0.7	−20	2.1	0.023	2.2	0.11	2.4	0.20	2.6	0.28	3.2	0.28	3.9	0.28
		+20	2.2	0.032	3.0	0.14	4.4	0.21	6.1	0.30	9.4	0.28	18.1	0.23
		50	2.3	0.036	3.2	0.14	4.5	0.20	6.1	0.28	9.4	0.26	18.1	0.22
		90	2.4	0.041	3.4	0.14	4.55	0.18	6.1	0.25	9.4	0.23	18.1	0.20
	0.8	−20	2.2	0.026	2.4	0.12	2.6	0.22	2.8	0.32	3.6	0.32	4.4	0.32
		+20	2.3	0.036	3.3	0.16	4.8	0.25	6.8	0.34	10.8	0.32	18.6	0.30
		50	2.4	0.040	3.5	0.16	4.9	0.24	6.8	0.31	10.8	0.30	18.6	0.27
		90	2.5	0.046	3.8	0.16	5.0	0.22	6.8	0.28	10.8	0.27	18.6	0.23
2.4×10^9	0.3	−20	1.3	0.011	1.5	0.05	1.6	0.08	1.7	0.12	2.0	0.12	2.3	0.12
		+20	1.4	0.015	1.8	0.06	2.1	0.10	2.7	0.13	4.1	0.13	7.8	0.11
		50	1.5	0.017	1.9	0.06	2.1	0.09	2.7	0.12	4.1	0.12	7.8	0.10
		90	1.55	0.019	2.05	0.06	2.2	0.08	2.7	0.10	4.1	0.10	7.8	0.09
	0.4	−20	1.5	0.014	1.7	0.06	1.8	0.11	1.9	0.16	2.3	0.16	2.7	0.16
		+20	1.6	0.019	2.0	0.08	2.5	0.14	3.2	0.18	5.3	0.15	10.2	0.15
		50	1.7	0.021	2.1	0.08	2.5	0.13	3.2	0.16	5.3	0.14	10.2	0.14
		90	1.8	0.024	2.3	0.08	2.6	0.11	3.2	0.14	5.3	0.13	10.2	0.12

Appendix 3 (Continued)

f (Hz)	Density in oven-dry condition	t (°C)	Moisture content (%)											
			0		10		20		30		60		100	
	ρ_0 (g/cm^3)		$\varepsilon'_\perp$	$tg\delta_\perp$	$\varepsilon'_\perp$	$tg\delta_\perp$	$\varepsilon'_\perp$	$tg\delta_\perp$	$\varepsilon'_\perp$	$tg\delta_\perp$	$\varepsilon'_\perp$	$tg\delta_\perp$	$\varepsilon'_\perp$	$tg\delta_\perp$
	0.5	−20	1.6	0.017	1.9	0.08	2.0	0.14	2.2	0.20	2.6	0.20	3.1	0.20
		+20	1.7	0.023	2.3	0.11	2.9	0.18	3.8	0.22	6.5	0.20	12.8	0.19
		50	1.8	0.026	2.4	0.11	2.9	0.17	3.8	0.20	6.5	0.18	12.8	0.17
		90	1.9	0.029	2.6	0.11	3.0	0.15	3.8	0.17	6.5	0.16	12.8	0.15
	0.6	−20	1.8	0.019	2.0	0.09	2.2	0.17	2.4	0.24	2.9	0.24	3.5	0.24
		+20	1.9	0.027	2.5	0.13	3.3	0.21	4.3	0.27	7.6	0.24	15	0.23
		50	2.0	0.030	2.7	0.13	3.4	0.19	4.3	0.24	7.6	0.22	15	0.21
		90	2.1	0.035	2.85	0.13	3.4	0.17	4.3	0.21	7.6	0.19	15	0.18
	0.7	−20	2.0	0.022	2.2	0.11	2.4	0.20	2.6	0.28	3.2	0.28	3.9	0.28
		+20	2.1	0.030	2.8	0.15	3.7	0.25	4.9	0.31	8.9	0.28	17.5	0.26
		50	2.2	0.034	3.0	0.15	3.8	0.23	4.9	0.28	8.9	0.25	17.5	0.24
		90	2.3	0.038	3.2	0.15	3.8	0.20	4.9	0.24	8.9	0.22	17.5	0.20
	0.8	−20	2.1	0.024	2.4	0.12	2.6	0.22	2.8	0.32	3.6	0.32	4.0	0.32
		+20	2.2	0.033	3.0	0.17	4.0	0.28	5.4	0.35	10.1	0.31	18	0.30
		50	2.3	0.037	3.2	0.17	4.1	0.25	5.4	0.32	10.1	0.28	18	0.27
		90	2.4	0.042	3.4	0.17	4.1	0.23	5.4	0.27	10.1	0.24	18	0.23
5.8×10^9	0.3	−20	1.3	0.010	1.5	0.04	1.5	0.07	1.6	0.10	1.9	0.10	2.2	0.10
		+20	1.4	0.014	1.7	0.06	2.0	0.12	2.7	0.16	3.7	0.15	7.1	0.15
		50	1.4	0.016	1.8	0.06	2.1	0.11	2.8	0.15	3.7	0.14	7.1	0.14
		90	1.5	0.019	1.9	0.06	2.1	0.10	2.9	0.13	3.7	0.12	7.1	0.12

	0.4	−20	1.4	0.013	1.6	0.05	1.7	0.09	1.8	0.13	2.2	0.13	2.6	0.13
		+20	1.5	0.018	1.9	0.075	2.3	0.16	3.3	0.21	4.8	0.20	9.3	0.20
		50	1.55	0.021	2.0	0.08	2.4	0.15	3.4	0.19	4.8	0.19	9.3	0.19
		90	1.6	0.024	2.2	0.08	2.5	0.13	3.5	0.16	4.8	0.16	9.3	0.16
	0.5	−20	1.6	0.016	1.8	0.07	1.9	0.12	2.1	0.17	2.5	0.17	3.1	0.17
		+20	1.7	0.022	2.15	0.10	2.6	0.21	3.8	0.26	5.9	0.25	11.6	0.25
		50	1.75	0.025	2.3	0.10	2.7	0.19	3.9	0.24	5.9	0.24	11.6	0.24
		90	1.8	0.030	2.5	0.10	2.8	0.17	4.1	0.20	5.9	0.20	11.6	0.21
	0.6	−20	1.7	0.019	2.0	0.08	2.1	0.14	2.3	0.20	2.8	0.20	3.4	0.20
		+20	1.8	0.026	2.3	0.12	3.0	0.25	4.4	0.32	6.9	0.31	13.6	0.31
		50	1.85	0.030	2.4	0.12	3.1	0.23	4.5	0.29	6.9	0.29	13.6	0.30
		90	1.9	0.035	2.6	0.12	3.2	0.20	4.7	0.25	6.9	0.25	13.6	0.26
	0.7	−20	1.9	0.021	2.1	0.09	2.3	0.16	2.5	0.23	3.1	0.23	3.8	0.23
		+20	2.0	0.029	2.5	0.14	3.4	0.29	4.9	0.37	8.1	0.36	15.8	0.36
		50	2.1	0.033	2.7	0.14	3.5	0.27	5.1	0.34	8.1	0.34	15.8	0.33
		90	2.1	0.039	2.9	0.14	3.6	0.24	5.2	0.29	8.1	0.29	15.8	0.28
	0.8	−20	2.1	0.023	2.3	0.10	2.4	0.18	2.7	0.26	3.4	0.26	4.2	0.26
		+20	2.2	0.032	2.7	0.16	3.6	0.33	5.5	0.43	9.2	0.41	16.2	0.41
		50	2.3	0.037	2.9	0.16	3.7	0.31	5.7	0.39	9.2	0.37	16.2	0.37
		90	2.35	0.043	3.1	0.16	3.85	0.27	5.9	0.34	9.2	0.32	16.2	0.32
10^{10}	0.3	−20	1.3	0.010	1.5	0.03	1.5	0.05	1.6	0.08	1.8	0.08	2.2	0.08
		+20	1.4	0.014	1.6	0.05	1.8	0.14	2.4	0.19	3.3	0.19	6.3	0.21
		50	1.4	0.017	1.7	0.05	1.9	0.13	2.5	0.17	3.5	0.17	6.7	0.19
		90	1.5	0.020	1.8	0.05	2.05	0.12	2.7	0.15	3.8	0.15	7.2	0.17
	0.4	−20	1.4	0.013	1.6	0.04	1.7	0.07	1.8	0.10	2.1	0.10	2.6	0.10
		+20	1.5	0.018	1.8	0.07	2.1	0.19	2.9	0.25	4.3	0.26	8.3	0.27
		50	1.55	0.021	1.9	0.07	2.2	0.18	3.1	0.23	4.6	0.24	8.8	0.25
		90	1.6	0.026	2.1	0.07	2.4	0.16	3.3	0.20	4.9	0.21	9.5	0.21

Appendix 3 (Continued)

f (Hz)	Density in oven-dry condition ρ_0 (g/cm³)	t (°C)	Moisture content (%) 0 $\varepsilon'_\perp$	0 $tg\delta_\perp$	10 $\varepsilon'_\perp$	10 $tg\delta_\perp$	20 $\varepsilon'_\perp$	20 $tg\delta_\perp$	30 $\varepsilon'_\perp$	30 $tg\delta_\perp$	60 $\varepsilon'_\perp$	60 $tg\delta_\perp$	100 $\varepsilon'_\perp$	100 $tg\delta_\perp$
	0.5	−20	1.6	0.016	1.8	0.05	1.9	0.09	2.0	0.13	2.4	0.13	3.0	0.13
		+20	1.7	0.022	2.0	0.09	2.4	0.24	3.4	0.31	5.3	0.31	10.3	0.34
		50	1.75	0.026	2.1	0.09	2.5	0.23	3.6	0.28	5.6	0.28	10.9	0.31
		90	1.8	0.031	2.3	0.09	2.7	0.21	3.9	0.24	6.0	0.24	11.7	0.27
	0.6	−20	1.7	0.019	1.9	0.06	2.0	0.11	2.2	0.16	2.7	0.16	3.3	0.16
		+20	1.8	0.026	2.1	0.10	2.7	0.28	3.8	0.37	6.2	0.38	12.2	0.41
		50	1.85	0.031	2.2	0.10	2.9	0.26	4.0	0.34	6.6	0.35	12.9	0.37
		90	1.9	0.037	2.4	0.10	3.1	0.24	4.3	0.29	7.1	0.30	13.9	0.32
	0.7	−20	1.9	0.021	2.1	0.07	2.2	0.13	2.4	0.18	3.0	0.18	3.7	0.18
		+20	2.0	0.029	2.3	0.12	3.0	0.33	4.3	0.43	7.2	0.44	14.1	0.47
		50	2.1	0.034	2.4	0.12	3.2	0.31	4.6	0.39	7.6	0.40	14.9	0.43
		90	2.1	0.041	2.6	0.12	3.4	0.28	4.9	0.34	8.2	0.35	16.1	0.37
	0.8	−20	2.1	0.023	2.2	0.8	2.4	0.14	2.6	0.21	3.2	0.21	4.1	0.21
		+20	2.2	0.032	2.5	0.14	3.2	0.38	4.8	0.5	8.2	0.51	14.5	0.53
		50	2.3	0.038	2.65	0.14	3.4	0.36	5.1	0.46	8.7	0.46	15.4	0.48
		90	2.35	0.045	2.85	0.14	3.6	0.33	5.5	0.40	9.3	0.40	16.4	0.42

Appendix 4

Dielectric properties, $\varepsilon'_{\perp}$ and $tg\delta_{\perp}$, of boards made from unbleached pine wood pulp, at frequencies of $10^6 - 4 \times 10^7$ Hz, ash content 0.28–0.43%, and with vector $\bar{E}$ oriented perpendicular to the sheet surface (Peskovets 1975)

f (Hz)	ρ_0 (g/cm^3)	t (°C)	Moisture content (%)			
			0		6.4	
			$\varepsilon'_{\perp}$	$tg\,\delta_{\perp}$	$\varepsilon'_{\perp}$	$tg\,\delta_{\perp}$
10^6	0.6	25	1.95	0.022	1.97	0.022
		55	1.94	0.014	1.96	0.016
		95	1.93	0.007	1.92	0.012
	0.8	25	2.36	0.025	2.38	0.026
		55	2.35	0.016	2.36	0.019
		95	2.33	0.008	2.33	0.015
	1.0	25	2.86	0.028	2.88	0.029
		55	2.85	0.018	2.86	0.022
		95	2.82	0.009	2.83	0.018
	1.2	25	3.55	0.031	3.57	0.033
		55	3.53	0.020	3.55	0.026
		95	3.50	0.010	3.52	0.022
10^7	0.6	25	1.93	0.030	1.96	0.029
		55	1.93	0.020	1.94	0.019
		95	1.91	0.010	1.91	0.010
	0.8	25	2.33	0.035	2.35	0.033
		55	2.32	0.023	2.34	0.022
		95	2.30	0.011	2.30	0.011
	1.0	25	2.82	0.039	2.85	0.037
		55	2.81	0.026	2.83	0.025
		95	2.79	0.012	2.79	0.013
	1.2	25	3.50	0.044	3.52	0.042
		55	3.48	0.029	3.50	0.028
		95	3.44	0.014	3.47	0.014
4.10^7	0.6	25	1.87	0.030	1.90	0.028
		55	1.86	0.029	1.88	0.027
		95	1.85	0.013	1.86	0.012
	0.8	25	2.24	0.035	2.28	0.033
		55	2.23	0.034	2.26	0.032
		95	2.22	0.015	2.22	0.014
	1.0	25	2.71	0.040	2.73	0.037
		55	2.69	0.039	2.71	0.036
		95	2.67	0.017	2.68	0.016
	1.2	25	3.34	0.045	3.35	0.042
		55	3.31	0.041	3.33	0.041
		95	3.28	0.019	3.30	0.018

Appendix 5

Dielectric properties of some cellulose and paper products with the electric strength vector oriented parallel to the paper sheet at a frequency of 10 GHz (Mikaelan 1978, 1982)

W (%)	t (°C)	Purified wood sulfite cellulose (α-cellulose not less than 98%, ash not exceeding 0.25%, resins and fats not more than 0.6%)		Paper for electrical engineering purposes (prehydrolysis cord 100% sulfate cellulose, iron up to 40 mg/kg, calcium up to 60 mg/kg, mass 30 g/m^2, density 0.8 g/cm^3, thickness 0.04 mm)		Paper for decorative coating (50% of coniferous and 50% of angiosperm cellulose, ash content 7%, mass 165 g/m^2, thickness 0.18–0.21 mm)	
		$\varepsilon'_{\parallel}$	$tg\delta_{\parallel}$	$\varepsilon'_{\parallel}$	$tg\delta_{\parallel}$	$\varepsilon'_{\parallel}$	$tg\delta_{\parallel}$
0	30			2.4		3.5	
	50			3.6		4.6	
	90			6.0	0.39	5.7	0.17
10	30	4.4	0.33	6.6	0.55	7.3	0.34
	50	5.0	0.30	8.1	0.49	8.3	0.28
	90	5.9	0.22	11.5	0.35	9.6	0.17
20	30	6.0	0.33	10	0.53	9.6	0.31
	50	6.5	0.28	11.9	0.46	10.3	0.25
	90	7.5	0.18	14.5	0.32	11.3	0.15
40	30	8.3	0.29	14.9	0.46	12	0.24
	50	8.9	0.23	16.2	0.40	12.6	0.18
	90	10.0	0.13	18.6	0.26	13.7	0.10
60	30	9.8	0.24	18.0	0.37	13.5	0.17
	50	10.4	0.19	19.2	0.33	14	0.12
	90	11.2	0.10	21.0	0.21	15	0.05
80	30	10.5	0.18			14	0.11
	50	11.1	0.14			14.7	0.08
	[illegible]	[illegible]	[illegible]			[illegible]	[illegible]

Appendix 6

Dielectric constant ($\varepsilon'_{\perp}$) and loss tangent ($tg\delta_{\perp}$) values of hardboards with density $\rho = 0.9$–$1.0\ g/cm^3$. Vector $\bar{E}$ is oriented perpendicular to the sheet surface (James 1975)

W(%)	t (°C)	Frequency (Hz)									
		10^2		10^3		10^4		10^6		5×10^7	
		$\varepsilon'_{\perp}$	$tg\delta_{\perp}$	$\varepsilon'_{\perp}$	$tg\delta_{\perp}$	$\varepsilon'_{\perp}$	$tg\delta_{\perp}$	$\varepsilon'_{\perp}$	$tg\delta_{\perp}$	$\varepsilon'_{\perp}$	$tg\delta_{\perp}$
Douglas-fir, wet process											
0	25	3.6	0.0075	3.2	0.016	3.0	0.026	3.2	0.047	2.9	0.078
	45	3.6	0.024	3.3	0.013	3.1	0.018	3.4	0.038	3.1	0.077
	90	4.1	0.051	3.7	0.030	3.5	0.018	3.6	0.028	3.3	0.085
4	25	4.3	0.028	4.2	0.017	4.1	0.018	3.8	0.054	3.4	0.094
	45	4.5	0.069	4.3	0.023	4.2	0.016	4.0	0.040	3.5	0.10
	90	5.6	0.53	4.7	0.13	4.4	0.036	5.6	0.025	5.1	0.075
8.3	25	8.3	0.81	6.1	0.24	5.5	0.07	5.0	0.047	4.4	0.11
	45	13.0	1.9	7.3	0.60	5.9	0.16	5.2	0.043	4.6	0.10
	90	24.0	14.0	13.0	3.0	7.0	0.76	6.3	0.068	5.1	0.084
11.6	25	36	4.6	15	1.6	9.0	0.48	6.1	0.087	5.2	0.12
	45	52	9.7	22	2.7	10	1.0	6.3	0.12	5.4	0.11
	90	85	64	40	14.0	19	3.4	7.6	0.37	5.9	0.095
17	25	120	17	40	6.0	18	1.7	8.0	0.20	6.4	0.13
	45	150	37	61	9.5	24	3.0	8.5	0.32	6.5	0.12
	90	460	74	140	25	53	6.6	12	0.88	7.0	0.17

Appendix 6 (Continued)

W(%)	t (°C)	Frequency (Hz)									
		10^2		10^3		10^4		10^6		5×10^7	
		$\varepsilon'_\perp$	$tg\delta_\perp$	$\varepsilon'_\perp$	$tg\delta_\perp$	$\varepsilon'_\perp$	$tg\delta_\perp$	$\varepsilon'_\perp$	$tg\delta_\perp$	$\varepsilon'_\perp$	$tg\delta_\perp$
White oak, dry process											
0	25	3.9	0.026	3.5	0.030	3.3	0.038	3.4	0.048	3.1	0.071
	45	3.8	0.026	3.6	0.028	3.3	0.032	3.6	0.042	3.2	0.071
	90	4.6	0.050	4.1	0.037	3.8	0.031	3.8	0.038	3.5	0.080
3.5	25	4.9	0.062	4.7	0.034	4.6	0.032	4.1	0.059	3.6	0.083
	45	5.2	0.096	4.9	0.048	4.7	0.057	4.4	0.046	3.8	0.096
	90	8.5	0.72	5.9	0.23	5.2	0.084	6.6	0.040	5.9	0.076
8.0	25	8.9	0.49	6.9	0.20	6.1	0.089	5.3	0.060	4.6	0.10
	45	15	1.4	8.9	0.51	6.8	0.19	5.6	0.063	4.8	0.10
	90	44	30	18	7.1	10	1.5	7.2	0.21	5.5	0.12
11.6	25	47	6.6	21	1.9	11	0.65	6.9	0.13	5.6	0.12
	45	82	17	33	4.3	14	1.4	7.4	0.20	5.8	0.13
	90	220	110	58	40	29	8.2	10	0.62	6.7	0.15
17.8	25	440	20	95	10	34	3.4	11	0.40	7.4	0.16
	45	330	83	120	24	44	7.0	12	0.62	7.5	0.18
	90	1200	130	160	100	66	25	19	1.4	8.4	0.30

Appendix 7

Dielectric properties of some plastic materials containing wood, cellulose, or paper (von Hippel 1954)

Description of the material	Composition	t (°C)	Dielectric properties	Frequency (Hz)						
				10^2	10^4	10^6	10^7	10^8	3×10^9	10^{10}
Bakelite BM-120 (preformed and preheated)	Phenol-formaldehyde resin 46%, wood flour 46%, additives 8%	25	ε' $tg\delta$	4.87 0.030	4.62 0.020	4.36 0.028	4.16 0.035	3.95 0.038	3.70 0.044	3.68 0.041
Formica XX ($\bar{E}_\perp$ to laminate)	Phenol-formaldehyde resin 50%, paper 50%, laminate	26	ε' $tg\delta$	5.23 0.023	4.96 0.017	4.60 0.034	4.32 0.049	4.04 0.057	3.57 0.060	3.55 0.070
Micarta 254	Cresylic acid formaldehyde 50%, α-cellulose 50%	25	ε' $tg\delta$	5.3 0.076	4.81 0.026	4.51 0.036	4.20 0.047	3.85 0.055	3.43 0.051	3.25 0.041
		82	ε' $tg\delta$	9.2 0.410	5.5 0.069	5.1 0.030	4.9 0.040	– –	4.02 0.098	3.21 0.077
Melmac molding compound 1500	Melamine-formaldehyde resin 42%, wood flour 40%, plasticizer 18%	25	ε' $tg\delta$	6.50 0.031	6.18 0.016	5.85 0.032	5.53 0.042	5.10 0.050	4.20 0.052	4.10 0.051
Resimene 803-A (not preformed or preheated)	Melamine-formaldehyde resin 60%, cellulose 40%	24	ε' $tg\delta$	7.05 0.041	6.75 0.019	6.20 0.044	5.70 0.066	5.20 0.082	4.53 0.082	4.23 0.076
Plaskon urea, brown	Urea-formaldehyde resin 72%, paper 28%	24	ε' $tg\delta$	6.65 0.032	6.15 0.024	5.75 0.028	5.42 0.035	5.0 0.042	4.6 0.053	4.55 0.074

Appendix 8

Density of oven-dry wood of some industrially important tree species

English name	Latin name	Density ρ_0 (g/cm^3)
8.1. Europe and North Asia		
Hardwoods		
Alder	*Alnus glutinosa*	0.50
Ash, European	*Fraxinus excelsior*	0.63
Ash, Japanese	*F. mandshurica*	0.62
Aspen, European	*Populus tremula*	0.46
Beech, European	*Fagus sylvatica*	0.66
Birch, Asiatic	*B. costata*	0.65
Birch, Silver	*Betula verrucosa, B. pubescens*	0.61
Elm	*Ulmus loevis*	0.51
Hornbeam, European	*Carpinus betulus*	0.76
Lime	*Tilia cordata*	0.46
Maple, Norway	*Acer platanoides*	0.67
Oak, English	*Quercus robur*	0.68
Poplar, white	*Populus alba*	0.39
Softwoods		
Fir, European Silver	*Abies alba*	0.40
Fir, Siberian	*A. sibirica*	0.36
Larch, European	*Larix decidua*	0.56
Larch, Siberian	*L. sibirica*	0.64
Pine, Korean	*Pinus koraiensis*	0.41
Pine, Scots	*Pinus sylvestris*	0.48
Pine, Siberian Yellow	*P. sibirica*	0.42
Spruce, Norway	*P. abies*	0.45
Spruce, Siberian	*Picea obovata*	0.42
8.2. North America		
Hardwoods		
Ash, white	*Fraxinus americana*	0.64
Aspen, quaking	*Populus tremuloides*	0.41
Basswood, American	*Tilia americana*	0.37
Beech, American	*Fagus grandifolia*	0.65
Birch, paper	*Betula papyrifera*	0.56
Birch, yellow	*B. alleghaniensis*	0.64
Chestnut, American	*Castanea dentata*	0.47
Eastern cottonwood	*Populus deltoides*	0.43
Elm, American	*Ulmus americana*	0.54
Hickory (mockernut)	*Carya tomentosa*	0.75

Appendix 8 (Continued)

English name	Latin name	Density ρ_0 (g/cm^3)
Magnolia, Southern	*Magnolia grandiflora*	0.54
Maple, sugar	*Acer saccharum*	0.61
Oak, northern red	*Quercus rubra*	0.65
Oak, white	*Q. alba*	0.70
Poplar, yellow, tulip tree	*Liriodendron tulipifera*	0.47
Softwoods		
Cedar, Port-Orford	*Chamaecyparis lawsoniana*	0.45
Cedar, western red	*Thuja plicata*	0.36
Douglas-fir	*Pseudotsuga menziesii*	0.52
Fir, balsam	*A. balsamea*	0.40
Fir, subalpine	*Abies lasiocarpa*	0.36
Hemlock, eastern	*Tsuga canadensis*	0.45
Hemlock, western	*T. heterophylla*	0.49
Larch, eastern tamarack	*Larix laricina*	0.58
Larch, western	*L. occidentalis*	0.56
Pine, jack	*Pinus banksiana*	0.47
Pine, pitch	*P. rigida*	0.63
Pine, ponderosa	*P. ponderosa*	0.44
Pine, western white	*P. monticola*	0.40
Sequoia, giant	*Sequoiadendron giganteum*	0.44
Spruce, Engelmann	*Picea engelmannii*	0.39
Spruce, white	*Picea glauca*	0.43

8.3. Wood Species Imported by the USA

From Africa		
Afara, limba	*Terminalia superba*	0.47
Afrormosia, kokrodua	*Afrormosia elata*	0.77
Afzelia	*Afzelia africana*	0.79
Albizia, West African	*Albizia*	0.61
Avodiré	*Turraeanthus africanus*	0.49
Berlinia	*Tetraberlinia tubmaniana*	0.63
Gaboon, okoumé	*Aucoumea klaineana*	0.35
Ilomba, akomu	*Pycnanthus angolensis*	0.42
Mahogany, African	*Khaya ivorensis*	0.50
Obeche, warva	*Triplochiton scleroxylon*	0.39
Sapele	*Entandrophragma cylindricum*	0.70
From Central and South America		
Balsa	*Ochroma pyramidale*	0.16
Banak	*Virola koschnyi*	0.51
Cativo	*Prioria copaifera*	0.47
Cedar, Central American	*Cedrela odorata*	0.48
Courbaril	*Hymenaea courbaril*	0.85
Degame	*Calyeophyllum candidissimum*	0.78

Appendix 8 (Continued)

English name	Latin name	Density ρ_0 (g/cm^3)
Jacareúba	*Callophyllum brasiliense*	0.69
Lignum-vitae	*Guiacum officinale*	1.04
Mahogany, Central American	*Swietenia macrophylla*	0.52
Oaks, Latin American	*Quercus costaricensis*	0.65
Parañá pine	*Araucaria angustifolia*	0.54
Pine, Caribbean	*Pinus caribaea*	0.79
Rosewood, Brazilian	*Dalbergia nigra*	0.83
From South-East Asia and Australia		
Apitong	*Dipterocarpus* spp.	0.69
Jarrah	*Eucalyptus marginata*	0.79
Kapur	*Dryobalanops* spp.	0.75
Keruing	*Dipterocarpus* spp.	0.70
Meranti	*Shorea* spp.	0.4–0.7
Mersawa	*Anisoptera* spp.	0.59
Ramin	*Gonystylys bancanus (macrophyllum)*	0.69
Red lauan, bataan	*Shorea negrosensis*	0.51
Rosewood, Indian	*Dalbergia latifolia*	0.88
Teak	*Tectona grandis*	0.66
Walnut, European	*Juglans regia*	0.55
8.4. Far East		
Akamatsu, Japanese red pine	*Pinus densiflora*	0.52
Ash, Japanese	*Fraxinus mandshurica*	0.55
Beech, Japanese	*Fagus crenata*	0.60
Hinoki	*Chamaecyparis obtusa*	0.40
Katsura	*Carcidiphylum japonicum*	0.47
Keyaki	*Zelkova serrata*	0.60
Larch, Japanese	*Larix kaempferi*	0.46
Makamba birch, Japanese birch	*Betula maximowicziana*	0.61
Oak, Japanese	*Quercus* spp.	0.45–0.90
Paulownia, kiri	*Paulownia tomentosa*	0.27
Pine, Korean	*Pinus koraiensis*	0.49
Spruce, Jezo, Hondo	*Picea jezoensis*	0.40
Sugi, Japanese cedar	*Cryptomeria japonica*	0.35

References

Akhadov JV 1972 Dielectric properties of pure liquids. State publishing house of standards, Moscow (in Russian)

Alekseev AA, Vjukov IE, Grabovsky VA, Naumov VV et al. (eds) 1965 Papermaker's guide, vol 2. Lesn Promst, Moscow (in Russian)

Auty RP, Cole RH 1952 Dielectric properties of ice and solid D_2o. J Chem Phys 20: 1309–1314

Beldi F, Balint J, Szabo J, Ruzsa B 1968 Die dielektrischen Eigenschaften verschiedener Eichenholzarten. Holz Roh Werkst 26: 89–95

Benkova VE 1981 Exploration of the properties of bound water in wood with the use of dielectric methods. Khin Drev 3: 96–98 (in Russian)

Benkova VE 1987 Dielectric relaxation of bound water in wood. Author's papers on Thesis for Candidate of Science degree. SIBTI, Krasnoyarsk (in Russian)

Benkova VE Chudinov BS 1990 Dielectric relaxation in moist wood. Khin Drev 1: 97–101 (in Russian)

Berdinskikh IP, Sikorskiy YA, Koblikova AG 1955 On wood dielectric constant values. Derevoobrab Promst 9: 16–17 (in Russian)

Bergmann K 1969 Dielektrische Messungen an festen Kunststaffen im Frequenzbereich 10^2–10^9 Hz. Z Ang Phys 28: 95–100

Birjukov VA 1964 Dependence of dielectric properties of wood on its density and on the alternating current frequency of electric field. Derevoobrab Promst 11: 13–15 (in Russian)

Boievskaja IA, Portnik ZhI, Meremjanin YI 1986 Influence of wood moisture content and density on its dielectric properties. Lesn Zh 1: 123–125 (in Russian)

Borovikov AM, Ugolev BN 1989 Wood guide. Lesn Promst, Moscow, (in Russian)

Brandt AA 1963 Investigation of dielectrics at super high frequencies. Fizmatizdat, Moscow (in Russian)

Breitweit KV, Koritskiy YV, Kulakova RV, Sokolova SL 1970 Production process, properties, and application of electrically insulating paper and board made of cellulose. Energia, Moscow (in Russian)

Brown GH, Hoyler CH, Bierwirth RA 1947 Theory and application of radio-frequency heating. Van Nostrand, New York, 264–269

Büchner A 1939 Das Mischkörpersproblem in der kondensatortechnik. Wiss Ver Siemens 84

Chudinov BS 1968 Theory of thermal treatment of wood. Nauka, Moscow (in Russian)

Chudinov BS 1984 Water in the wood. Nauka, Novosibirsk (in Russian)

Chudinov BS, Andreev MD 1978 Water in the cell walls. SO AN USSR, Krasnoyarsk, 15–17 (in Russian)

Cole KS, Cole RH 1941 Dispersion and absorbtion in dielectrics. I. Alternating current characteristics. J Chem Phys 9: 341–351

Dolgaleva AA 1958 Influence of the quality of water on the production process and on the properties of capacitor paper. Bum Promst 5: 10–13 (in Russian)

Efremov AN, Zhilenkov IV 1954 Dielectric parameters of compressed wood. VSKhI Proc Voronezh 25 (1): 225–230 (in Russian)

Eme F 1967 Dielectric measurements. Khimiya, Moscow, 108–110 (in Russian)

Flate DM 1986 Properties of paper. Lesn Promst, Moscow (in Russian)

Flate DM, Grunin YB 1973 Exploration of interactions between water and cellulose fibers with the use of nuclear magnetic relaxation. Bum Promst 10: 1–3 (in Russian)

Freidin AS 1961 Action of ionizing radiation on wood and on its components. Goslesbumizdat, Moscow (in Russian)
Glukhov VI, Matveev VK, Shiriajeva GV 1973 Dielectric properties of some plastics of wood origin. Plastics 5: 25–27 (in Russian)
Goriajev AA 1975 Exploration of dielectric method of drying spruce wood in vacuum. Author's Paper on Thesis for Candidate of Science Degree, MLTI, Moscow (in Russian)
Handa T, Fukuoka M, Yoshizawa S 1982 The effect of moisture on dielectric relaxation in wood. J Appl Polym Sci 27: 439–453
Hearmon RFS, Burcham JN 1954 The dielectric properties of wood. Her Majesty's Stationery Office, London, Spec Rep 8
Hermans PH 1949 Physics and chemistry of cellulose fibers. Elsevier, Amsterdam New York, 181–213
Hoffmann G, Poliszko S 1985 Zjaviska relaksacji dielektrycznej w drevnie buka (*Fagus silvatica* L) w przedziale temperatur od 100 K do 450 K (Dielectric relaxation in beech wood in temperature range from 100 K to 450 K). Folia For Pol 16: 5–14 (in Polish)
Ishida Y, Yoshino M, Takayanagi M, Irie F 1959 Dielectric studies of cellulose fibers. J Appl Polym Sci 1(2): 227–235
Ivanov AI 1984 Development of new methods of improving the electrically insulating properties of capacitor paper. Thesis for Candidate of Science Degree, LTI CBP, Leningrad (in Russian)
James WL 1975 Dielectric properties of wood and hardboard: variation with temperature, frequency, moisture content, and grain orientation. US For Serv Res Pap FPL-245
James WL, Hamill DW 1965 Dielectric properties of Douglas-fir measured at microwave frequencies. For Prod J 15: 51–56
Kasarnovskiy DM, Tareev BM 1980 Testing of electrically insulating materials and products. Energia, Leningrad, 17–95 (in Russian)
Kitajeva SKh 1960 Investigation on the influence of ash containing elements of wood sulphate cellulose on its dielectric losses. Author's paper on Thesis for Candidate of Science degree, LTA, Leningrad (in Russian)
Kitajeva SKh, Korolev VV 1957 Study of the electrodialysis process of cellulose. Bum Promst 6: 4–7 (in Russian)
Klotinj JE, Darsinsh TA 1983 Dielectric properties of wood impregnated with ammonia. In: Modification of properties of wood-based materials. Zinatne, Riga, 88–93 (in Russian)
Kolosovskaja EA, Loskutov SR, Chudinov BS 1989 Physical base of interactions between wood and water. Nauka, Novosibirsk (in Russian)
Kröner K, Pungs L 1953 Uber des Verhalten des dielektrischen Verlustfaktors von Naturholz im Groβen Frequenzbereich. Holzforschung 7: 12–18
Lee ST 1964 The dielectric properties of wood in the radio frequency region. Sci Silv 9(3): 223–245
Lichtenecker K 1926 Die Dielektrizitätskonstante natürlicher und künstlicher Mischkörper. Physik Z 27: 115–116
Maeda H, Fukada E 1987 Effect of bound water on piezoelectric, dielectric and elastic properties of wood. J Appl Polym Sci 33: 1187–1198
Maeda H, Tauda K, Fukada E 1977 Electrical and mechanical properties of bamboo and wood with various hydration. Rep Prog Pol Phys Jap 20: 739–742
Metaxas AC, Driscoll JL 1974 A comparison of the dielectric properties of paper and board at microwave and radio frequencies. J Microwave Power 9: 80–89
Miagkov IJ 1954 Experimental researches of electric properties of compressed wood of birch, lime, alder, aspen, and beech. Author's paper on Thesis for Candidate of Science degree, VISI, Voronezh (in Russian)

Mikaelan GG 1978 Frequency and moisture characteristics of dielectric parameters of paper in the super high frequency range. Bum Promst 5: 27–29 (in Russian)
Mikaelan GG 1982 Investigation and development of SHF method of checking the moisture profile of paper sheet for the purpose of improvement of paper quality. Author's paper on Thesis for Candidate of Science degree, LTI CBP, Leningrad, (in Russian)
Mikhailovskaja KP 1972 Investigation of moisture characteristics of wood electric parameters. Author's paper on Thesis for Candidate of Science degree, STI, Krasnoyarsk (in Russian)
Milov BG, Kitajeva SKh 1959 Diminution of dielectric losses in the cellulose intended for production of paper for capacitors and high voltage cables. Bum Promst 5: 4–7 (in Russian)
Milov BG, Kitajeva SKh, Bobrov AI, Waisman LM, Eremenko RK 1974 Electrically insulating paper. Lesn Promst, Moscow (in Russian)
Nanassy AJ 1970 Overlapping of dielectric relaxation spectra in oven-dry yellow birch at temperature from 20 to 100° C. Wood Sci Technol 4: 104–121
Nanassy AJ 1972 Dielectric measurement of moist wood in a sealed system. Wood Sci Technol 6: 67–77
Nekrasov LB 1979 Basic knowledge on electrical and mechanical destruction of frozen rock. Nauka, Novosibirsk, 47–74 (in Russian)
Netushil AV, Zhukhovitskiy BJ, Kudin VN, Parini EP 1959 High frequency heating of dielectrics and semiconductors. Gosenergoizdat, Moscow, 38–154 (in Russian)
Norimoto M 1976 Dielectric properties of wood. Wood Res 59/60: 106–152
Norimoto M, Hayashi S, Yamada T 1978 Anisotropy of dielectric constant in coniferous wood. Holzforschung 32: 167–172
Norimoto M, Yamada T 1972 The dielectric properties of wood VI. Wood Res 52: 31–43
Norimoto M, Yamada T 1975 Dielectric properties of cellulose irradiated with gamma-rays. Mokuzai Gakkaishi 21: 151–156
Olekhnovich AM, Tumanov SD, Chernonog YF 1974 Electric properties of modified birch wood. In: Batin NA (ed) Mechanical processing of wood, vol 4 Vyshejshaja Shkola, Minsk, 165–174 (in Russian)
Olekhnovich AM, Tumanov SD, Olekhnovich FM 1975 Electric properties of wood modified with polyethylene. In: Batin NA (ed) Mechanical processing of wood, vol 5. Vyshejshaja Shkola, Minsk, 162–171 (in Russian)
Olekhnovich AM, Turov LF, Olekhnovich FM 1976 Electrical and thermophysical properties of wood modified with formaldehyde resin. In: Batin NA (ed) Mechanical processing of wood. vol 6. Vyshejshaja Shkola, Minsk, 197–206 (in Russian)
Osaki S 1989 Dielectric anisotropy of nonwoven fabrics by using the microwave method. Tappi 72: 171–175
Otto SE 1968 Investigation of production process of particleboards manufacturing by the composite heating method. Thesis for Candidate of Science degree, MLTI, Moscow (in Russian)
Peskovets VI 1975 Exploration of electric properties of cellulose based materials in the high frequency range. Thesis for Candidate of Science degree STI, Krasnoyarsk (in Russian)
Peterson RW 1960 The dielectric Properties of wood. For Prod Lab Can, Ottawa, Tech Note 16
Peyskens E, de Pourcq M, Stevens M, Schalck J 1984 Dielectric properties of softwood species at microwave frequencies. Wood Sci Technol 18: 267–293
Postnikov LL 1972 Exploration of electrical and physical properties of chemically plasticized birch wood. Author's paper on Thesis for Candidate of Science degree, Riga (in Russian)

Pungs L 1954 Holz als Dielektrikum im Hochfrequenzfeld. Elektrotech Z A 75(13): 433–438

Pungs L, Lamberts K 1954 Untersuchungen über die Anwendung von Hochfrequenzerwärmung bei der Herstellung von Holzspanplatten. Holz Roh Werkst 12: 20–25

Rafalski J 1966 Über die dielektrischen Eigenschaften unterschiedlich verdichteten Rotbuchenvollholzes. Holztechnologie 7: 118–122

Reginač L, Makoviny I 1983 Dielectricke vlastnosti dreva. (Dielectric properties of wood.) Drev Vysk 2: 21–61 (in Czech)

Renne VT 1969 Electric capacitors. Energia, Leningrad, 356–372 (in Russian)

Rosenberg VI 1979 Scattering and attenuation of electromagnetic radiation by the atmospheric particles. Gydrometizdat, Leningrad (in Russian)

Rotsens KA, Berzon AV, Gulbis JK 1983 Specific characteristics of modified wood. Zinatne, Riga (in Russian)

Skaar C 1948 The dielectric properties of wood at several radio frequencies. NY State Coll For, Syracuse, NY, Tech Publ 69

Skaar C 1988 Wood-water relations. Springer, Heidelberg

Sobolev YS 1979 Wood as a structural material. Lesn Promst, Moscow, 9–10 (in Russian)

Takeda M 1951 Studies on dielectric behaviour of bound water in timber in the high frequency region. Bull Chem Soc Jap 24: 169–173

Tareev BM 1982 Physics of dielectric materials. Energoizdat, Moscow (in Russian)

Tinga WR 1969 Dielectric properties of Douglas-fir at 2.45 GHz. J Microwave Power 4: 162–164

Tiuri M, Jokela K, Heikkila S 1980 Microwave instrument for accurate moisture and density measurement of timber. J Microwave Power 15: 251–254

Torgovnikov GI 1982 Dielectric properties of basic wood species in the super high frequency range. Derevoobrab Promst 5: 4–6 (in Russian)

Trapp W, Pungs L 1956a Bestimmung der dielektrischen Werte von Cellulose, Glucose und der Zellsubstanz von Naturholz im großen Frequenzbereich. Holzforschung 10: 65–68

Trapp W, Pungs L 1956b Einfluß von Temperatur und Feuchte auf das dielektrische Verhalten von Naturholz im großen Frequenzbereich. Holzforschung 10: 144–150

Tsutsumi J 1967 Studies on dielectric properties of wood. Kyushu Univ For Bull 41: 109–169

Ugolev BN 1986 Wood Science and basic knowledge of wood commodity research. Lesn Promst, Moscow (in Russian)

Uyemura T 1960 Dielectric properties of wood as the indicator of the moisture. Gov For Exp Stn Tokio Bull 119: 55–172

Venkateswaran A 1972 A comparison of the electrical properties of milled wood, milled wood cellulose, and milled wood lignin. Wood Sci 4: 248–253

Venkateswaran A 1973 Effect of compression on electrical properties of wood and cellulose. Wood Sci 5: 230–234

Von Hippel AR 1954 Dielectric materials and applications. MIT, New York

Von Hippel AR 1959 Dielectrics and waves. John Wiley, New York

Zhilenkov IV 1957 Debye's dispersion of adsorbed water at low frequencies. Proc Acad Sci USSR Chem Sci Dept 2: 232–234 (in Russian)

Zhilenkov IV 1970 Dielectric properties of sorption structurized water. In: Proc 1st All-Union Conf Theory of adsorption. Nauka, Moscow, 235–243 (in Russian)

Subject Index

MIX
Papier aus verantwortungsvollen Quellen
Paper from responsible sources
FSC® C105338

If you have any concerns about our products,
you can contact us on
ProductSafety@springernature.com

In case Publisher is established outside the EU,
the EU authorized representative is:
Springer Nature Customer Service Center GmbH
Europaplatz 3, 69115 Heidelberg, Germany

Printed by Libri Plureos GmbH
in Hamburg, Germany